Electronics
for Electrical Installation

Electronics for Electrical Installation

R.D. Puckering

and

S. Brough

Department of Engineering and Building Services,
City of Westminster College, London

OXFORD
BLACKWELL SCIENTIFIC PUBLICATIONS
LONDON EDINBURGH BOSTON
MELBOURNE PARIS BERLIN VIENNA

Blackwell Scientific Publications
Editorial Offices:
Osney Mead, Oxford OX2 0EL
25 John Street, London WC1N 2BL
23 Ainslie Place, Edinburgh EH3 6AJ
3 Cambridge Center, Cambridge,
Massachusetts 02142, USA
54 University Street, Carlton
Victoria 3053, Australia

Other Editorial Offices:
Librairie Arnette SA
2, rue Casimir-Delavigne
75006 Paris
France

Blackwell Wissenschafts-Verlag
Meinekestrasse 4
D-1000 Berlin 15
Germany

Blackwell MZV
Feldgasse 13
A-1238 Wien
Austria

First published 1992

Set by DP Photosetting, Aylesbury, Bucks
Printed and bound in Great Britain by
Hartnolls, Bodmin, Cornwall

DISTRIBUTORS

Marston Book Services Ltd
PO Box 87
Oxford OX2 0DT
(*Orders:* Tel: 0865 791155
Fax: 0865 791927
Telex: 837515)

USA
Blackwell Scientific Publications, Inc.
3 Cambridge Center
Cambridge, MA 02142
(*Orders:* Tel: 800 759-6102
617 225-0401)

Canada
Oxford University Press
70 Wynford Drive
Don Mills
Ontario M3C 1J9
(*Orders:* Tel: 416 441-2941)

Australia
Blackwell Scientific Publications
(Australia) Pty Ltd
54 University Street
Carlton, Victoria 3053
(*Orders:* Tel: 03 347-0300)

British Library
Cataloguing in Publication Data

ISBN 0–632–03266–9

Library of Congress
Cataloging in Publication Data

Electronics for electrical installation /
R.D. Puckering and S. Brough.
p. cm.
Includes index.
ISBN 0–632–03266–9
1. Electronic circuits. 2. Electronics.
I. Puckering, R.D.
II. Brough, S.
TK7867.E424 1992
621.381′5—dc20 91-48039
CIP

Contents

Preface

In recent years we have seen an increase in the use of electronic devices and circuitry in electrical plant and equipment. Because of this employers in the electrical contracting industry have seen the need for electricians to have some basic knowledge of electronics, and this subject has been included as one of the units of the City and Guilds 236 Electrical Installation Course which leads to the National Vocational Qualification for the industry.

This new system of credit accumulation towards vocational qualifications has been introduced by the National Council for Vocational Qualifications (NCVQ) and is based on units of competence. Qualifications will be made up of a number of units which are recognisable, and have value, in a particular employment the electronics unit being one such unit.

Many of the tests carried out on electrical installations by electricians can in fact cause serious damage to certain electronic components. It is essential that the electrician be fully aware of this and is able to take the necessary action to prevent it happening.

Although it is not the intention of this book to turn electricians into electronic experts, it will give the students an awareness of the different electronic components together with their functions and principles of operation in electronic circuits. It will also show how to carry out simple fault diagnosis, tests on components and replacement of faulty parts.

The book will be suitable for those people taking the City and Guilds 236 Course, the Supplementary Studies in Electronics Certificate and the City and Guilds 232 Electrical and Electronic Craft Studies Course. The book will also provide practical help to people on BTEC electrical courses and the electrical installation part of building services courses.

Chapter 1
The Structure of Materials

1.1 The structure of the atom

Electronics derives its name from one of the constituent parts of the atom, namely the electron. Atoms, and the way they and their constituent parts behave, are of fundamental importance in the operation and behaviour of electronic components, materials and circuitry. It is because of this that no book on electronics would be complete without some reference, however simple, to the ultramicroscopic world of the atom.

The basic substances in our environment are known as elements and are classified according to their atomic structure. Each element has a structure which is unique to itself and all other substances are made up from combinations of two or more of these elements. Atoms are the smallest particles that can exist in nature without losing their identity as elements and can be regarded as the basic building blocks of nature.

Here are some basic facts about atoms:

- Virtually all matter in the universe is composed of atoms.
- Atoms of different elements vary in size, density and construction.
- There are about 100 different sorts of atom, each of which is a different chemical element such as hydrogen, oxygen, carbon or copper.
- All are too small to be seen even with the most powerful of microscopes.
- Each atom is composed of a nucleus and one or more electrons.
- The nucleus is composed of protons and neutrons.
- The protons each carry a positive charge of electricity and the neutrons carry no charge.
- The electrons which are orbiting the nucleus carry a negative charge of equal and opposite value.

There are 92 naturally occurring elements in all and they may be arranged according to the number of electrons their respective atoms contain, in ascending order. In this way an element can be identified in terms of its atomic number.

The hydrogen atom has one orbiting electron and one proton; hence its atomic number is 1. Next is helium which has two orbiting electrons, two protons and two neutrons, and its atomic number, as you would expect, is 2. As the atomic number increases so the structure becomes more complex (see Fig. 1.1). It is popular to picture the electrons moving round the nucleus in much the same way as the

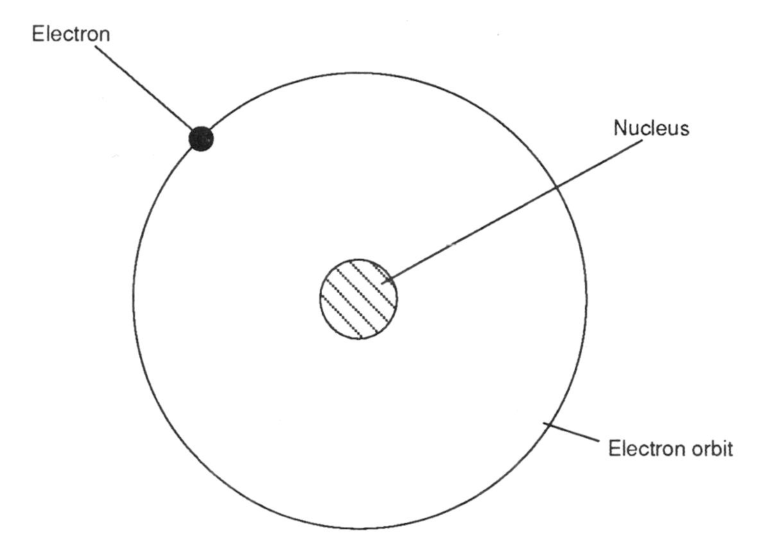

Fig. 1.1 Hydrogen atom.

planets encircle the sun in our solar system. They orbit in more than one plane (see Fig. 1.2) and some electrons move closer to the nucleus than others.

1.2 Electron flow

When the number of electrons orbiting the nucleus equals the number of protons in the nucleus the atom is said to be electrically neutral. Figure 1.2 shows four protons in the nucleus and four electrons orbiting, and the atom can be said to be in electrical balance.

Electrons which orbit furthermost from the nucleus are not so firmly attached as those that are closer, and these may more easily become detached or forced away from the nucleus. Materials which have loosely attached electrons are mostly metals and are good conductors of electricity. Materials whose electrons are more firmly attached and therefore more difficult to force away from the nucleus are things like mica and plastics which act as good insulators.

If for some reason an atom loses an electron it becomes positively charged and is known as a positive ion (see Fig. 1.3).

If the atom gains an electron it becomes negatively charged and is known as a negative ion. Since the atom is no longer neutral under these conditions it could be looked upon as a charged particle.

Any atom that has lost electrons becomes positively charged and strives constantly to achieve electrical balance by attracting loose negative electrons. The movement of electrons that takes place as a result can be described in simple terms as an electron flow. We shall see later that this flow is actually in the opposite direction to conventional current flow.

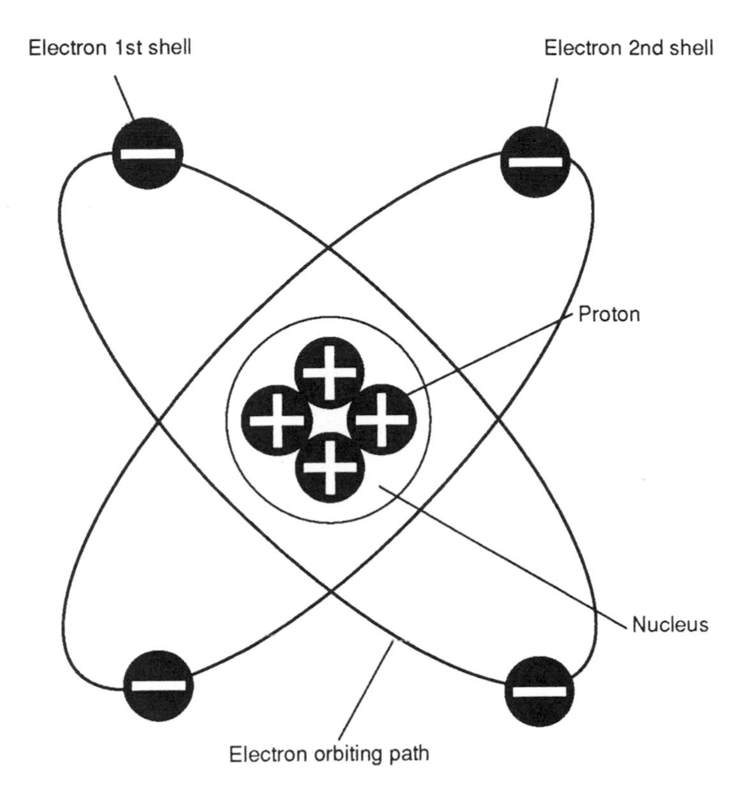

Fig. 1.2 Electrons orbiting.

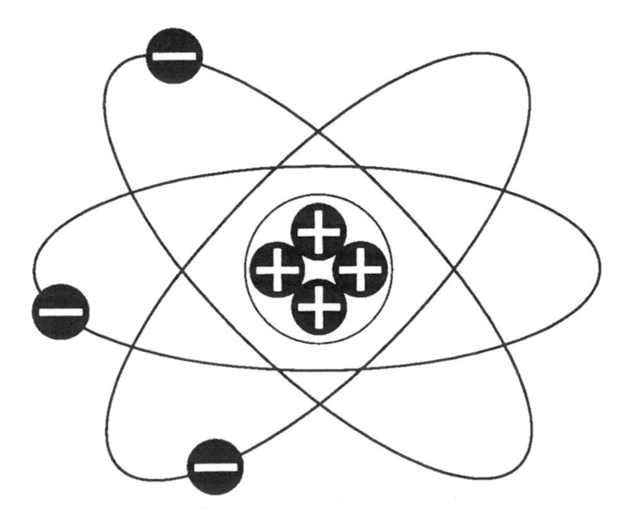

Fig. 1.3 Positively charged atom.

1.3 Production of electrical potential

It is a basic fact of physics that energy is not lost but is simply converted into some other form of energy. Events have to follow the laws of conservation of energy, so if we need to produce a particular form of energy for our own use we will have to expend other forms of energy to produce it.

In order to make full use of the phenomenon described above it would be necessary to detach electrons in order to create the unbalanced condition necessary to cause electrons to move and produce a current flow. To achieve this we need to apply some other form of energy.

There are a number of forms of energy that can be used to produce electron flow. It might be a good idea to examine in a simple way some of the ones that have particular importance to electronic devices used in the electrical industry.

Thermal energy

When two different metals – for example, copper and iron – are brought into contact with each other and heat is applied at the point of contact, then an electrical charge is produced. There are many different combinations of metals that can be used to produce this effect, though it would be foolish to utilize metals that would melt at the temperatures at which they are expected to be used.

We take advantage of the above fact in a device called a *thermocouple* where the two metals are welded together to form what is known as a junction. If the junction is connected to a sensitive meter (as shown in Fig. 1.4), and heat is applied, an electrical imbalance will result. The circuit will then act as a path for the current flow that is produced as a result of the electrons attempting to restore the electrical

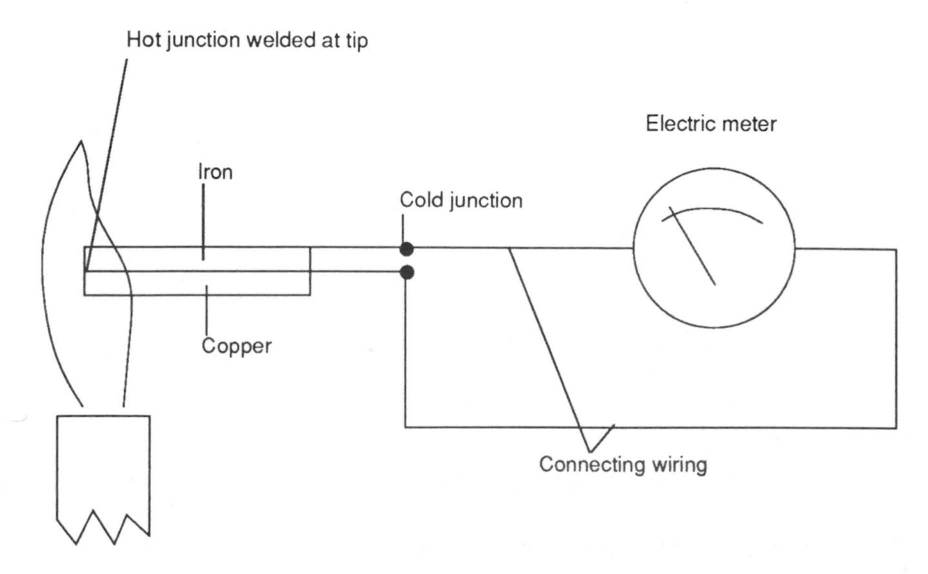

Fig. 1.4 Thermocouple.

balance to the atoms of the metal that has lost them. This flow of current will be indicated on the meter. The current is extremely small and the direction of flow is from negative to positive. The heated end of the junction is known as the *hot junction* and the other end of the device is known as the *cold junction*. The greater the temperature difference between the hot and cold junctions the greater the current available.

A practical application for the use of thermocouples is in the *thermoelectric flame failure devices* used in gas- or oil-fired boilers. Here a thermocouple is placed in the flame of the boiler and connected in such a way that if the flame goes out the electrical current produced by the device no longer flows and the gas or oil supply is automatically switched off.

Magnetic energy

We are aware, from our basic electrical theory, of the properties of permanent magnets; these properties are used to good effect in the electrical industry for numerous applications. For example, they are often employed to hold an appliance door shut, such as a refrigerator, and permanent magnets are used for some fractional kW motors, television scanning components, etc. How can we utilize these effects to produce electricity?

If we were to connect a centre scale galvanometer, as shown in (a) of Information Sheet No. 1A, and then move a section of the wire across the face of one pole of a permanent magnet, the needle of the galvanometer would move in one direction and then return to the central position. This indicates that a current was produced momentarily. Moving the wire across the face of the magnet in the other direction will cause the needle to move in the opposite direction and return to the central position, showing that a current flow occurred in the opposite direction.

Further experiments would show that:

- Holding the wire stationary produces no deflection of the meter and therefore no current flow.
- The direction in which the current flows depends upon which direction we make the wire cross the pole.
- If we use the opposite pole of the magnet the effect of the direction of current flow is reversed.
- The faster the wire is moved across the face of the magnetic pole the greater the deflection on the meter.
- The closer we pass the wire to the magnetic pole without actually touching it the greater the deflection.
- If we increase the strength of the permanent magnet we find the deflection is increased.
- When the wire is formed into a coil (see drawing (b) of Information Sheet No. 1A) the deflection increases.

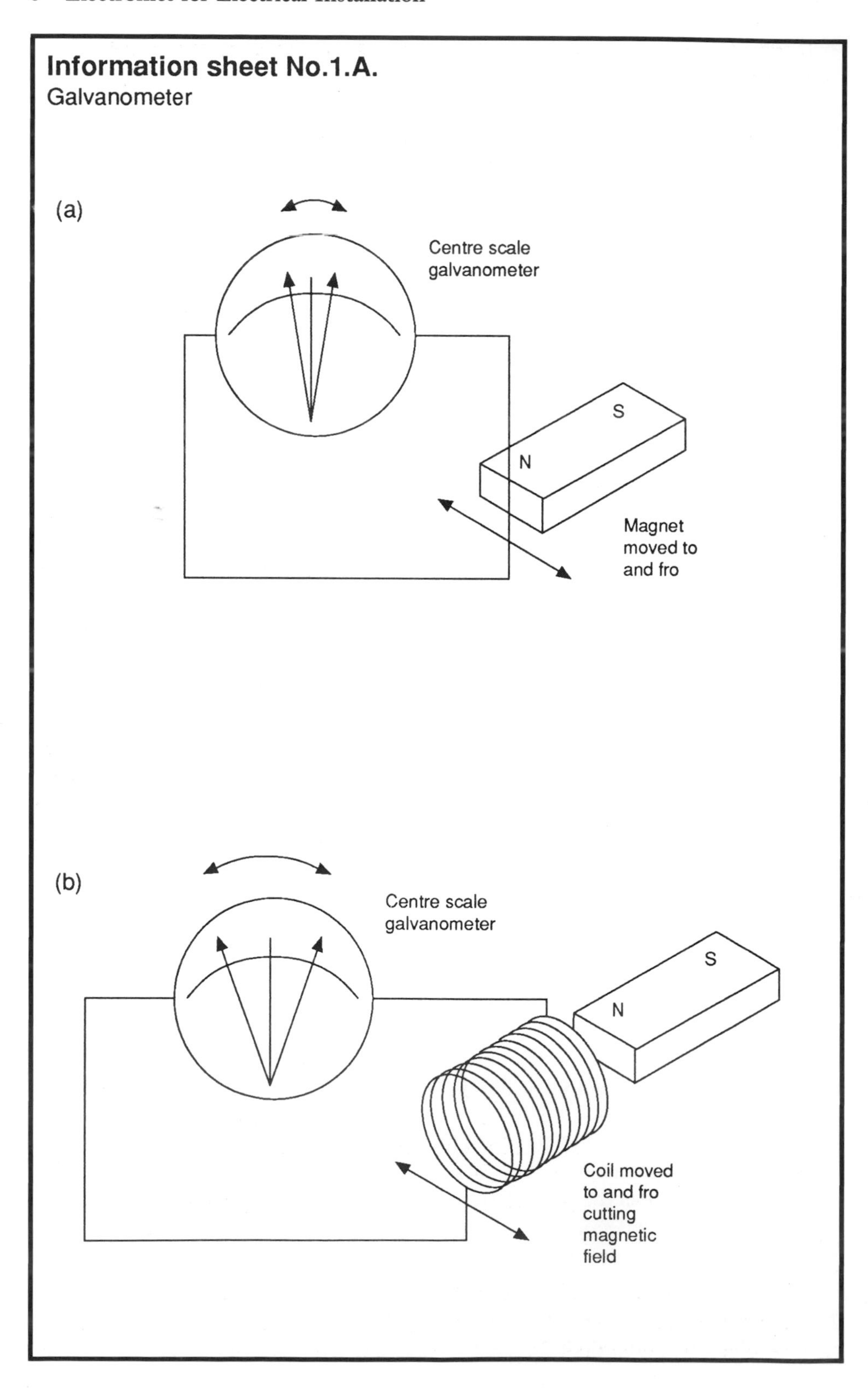
Information sheet No.1.A.
Galvanometer
(a)
Centre scale galvanometer
S
N
Magnet moved to and fro
(b)
Centre scale galvanometer
S
N
Coil moved to and fro cutting magnetic field

- If the number of coils passing the face of the magnetic pole is increased the amount of deflection increases.

In the above experiments the wire which is moved across the face of the magnetic pole is cutting the lines of magnetic force produced by the magnet, causing movement of electrons in the wire and producing a current flow. The more lines of force we can cut in a given time the greater the production of electricity and this effect is enhanced when we use a coil of wire instead of a single turn.

The above principles are used in the generation of electricity (see *Electrical Installation Practice, Book Three*) for our homes, factories, commercial premises, transportation, etc.

Chemical energy

When two different metals - for example, copper and iron - are placed in an electolyte, a flow of electrons takes place. There are many different combinations of metals that can be used to produce this effect, and indeed carbon can also be used.

We take advantage of the above fact in a device called a *simple cell* where the two metals, usually zinc and copper, are immersed in an electrolyte such as dilute sulphuric acid. The electrolyte acts chemically on the zinc and a difference in potential is produced between the two metals. If the two metals, known as electrodes, are connected to an external circuit, current will flow from the copper through the circuit to the zinc electrode and through the cell from zinc to copper.

The production of the electrical energy eventually causes the chemicals to weaken and break down and this, as well as problems with hydrogen bubbles forming on the copper electrode, causes the electron flow to cease (see Information Sheet No. 1B).

1.4 Semiconductor materials

A simple definition of a semiconductor is that its conductivity lies somewhere between a perfect insulator and a perfect conductor. The most widely used semiconductor materials in electronics are silicon and germanium, but cadmium sulphide is also extensively used for photoelectric devices.

Due to the atomic structure of germanium and silicon, they only behave like perfect insulators at absolute zero, with silicon having the better thermal stability. At room temperature, due to thermal agitation, the occasional electron is able to break loose.

The electrons in an atom's outer orbit, or shell as it is often called, are known as valence electrons. Each atom is arranged in a regular pattern with its four valence electrons interchanging with four valence electrons from four of its

Information sheet No.1.B.
Simple cell

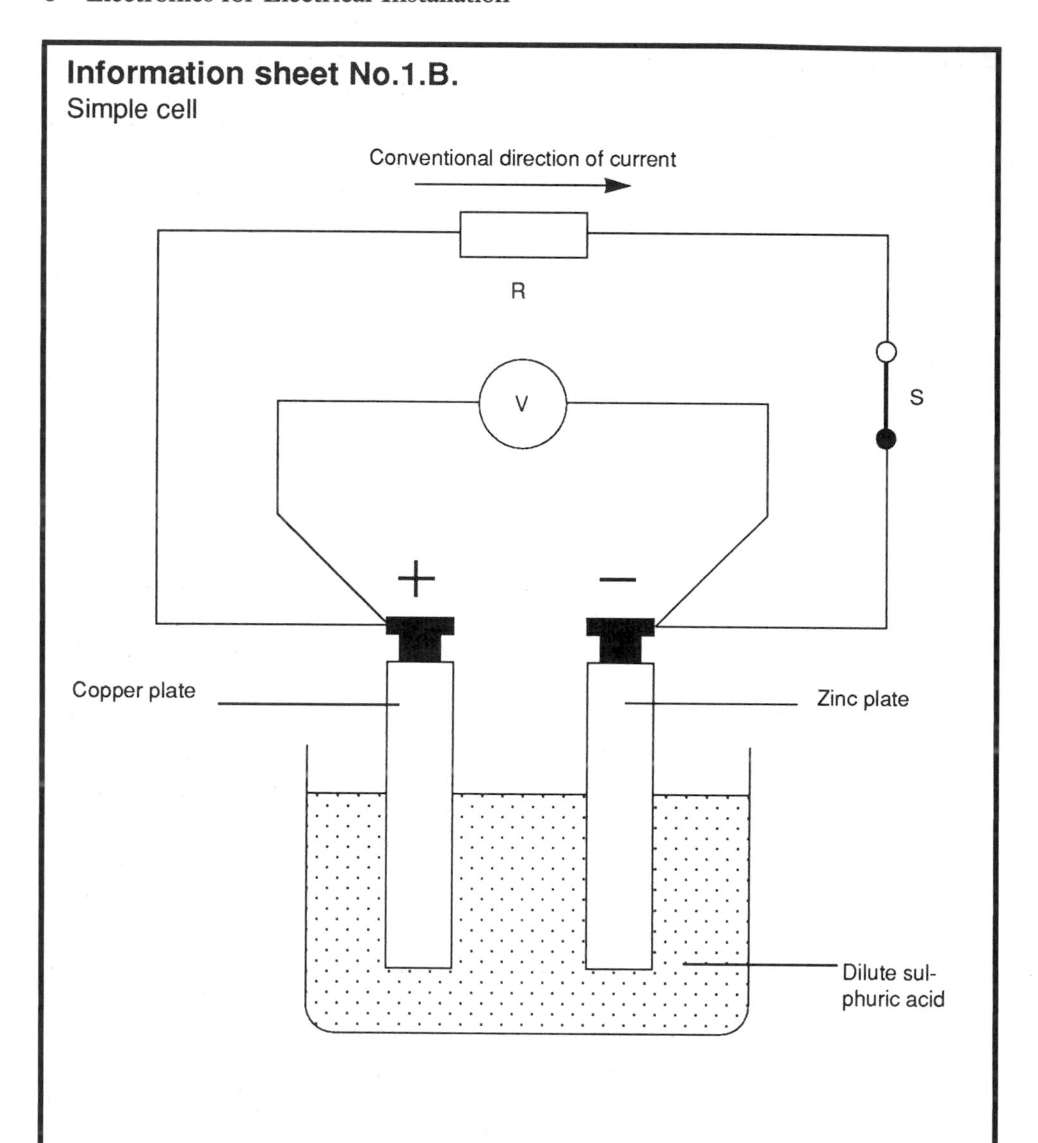

The drawing shows a resistor R and a switch S in series across copper and zinc electrodes immersed in dilute sulphuric acid.
Since the copper is positive relative to zinc electrode, current flows through R in the direction shown by the arrow.
This is conventional direction of current flow, actual electron flow is in the opposite direction.

neighbouring atoms in a lattice, as shown on (a) of Information Sheet No. 1C. The paths of interchange made by the electrons are called covalent bonds.

In order to improve the poor conductivity of the pure semiconductor material for use in electronic components such as diodes, transistors, etc., minute traces of impurity are added to it. The process of adding the impurities is called doping. Two groups of impurities are added; they are:

(1) *Trivalent impurities* – such as aluminium, gallium or indium – each containing three valence electrons.
(2) *Pentavalent impurities* – such as arsenic, antimony or phosphorus – each containing five valence electrons.

If the semiconductor material is doped with a trivalent impurity such as aluminium, which has three electrons in its outer shell, then, since the structure should have four, a deficiency or hole is introduced into the lattice. When trivalent impurities are added there is a tendency for the impurity atoms to 'steal' electrons from neighbouring semiconductor atoms in order to make up their deficiency. Since the electrons are negatively charged carriers, the region left behind is often referred to as a positively charged hole and the material is called p-type. When electrons move to fill a hole, they leave behind another hole, and thus it appears that the holes move. The impurity atoms in this type of material are called acceptors since they accept electrons from the surrounding atoms (see Information Sheet No.1C).

When a pentavalent material is added – e.g. phosphorus – which has five electrons in its outer shell, then, since only four electrons will fit into the structure, the fifth electron is spare and becomes free to act as a current carrier. A crystal treated in this way is called an n-type, because it contains free electrons which are negative charge carriers. The impurity atoms are called donor atoms since each donates a free electron to the lattice (see (c) on Information Sheet No.1C).

If, during manufacture, semiconductor material is doped (impurities are added) so as to produce p-type at one end and n-type at the other, a very important discovery is made in that a significant flow of current takes place in one direction only. This pn junction, as it is called, is at the heart of semiconductor devices and is discussed in detail in Chapter 3.

SUMMARY

This chapter has described in simple terms the structure of materials. The following important points should have been learned:

- Virtually all matter is composed of atoms.
- In materials that are good conductors the outer electrons are not very firmly attached and can be forced away from the nucleus.

Information sheet No.1.C.

Valence electrons

(a) Silicon lattice

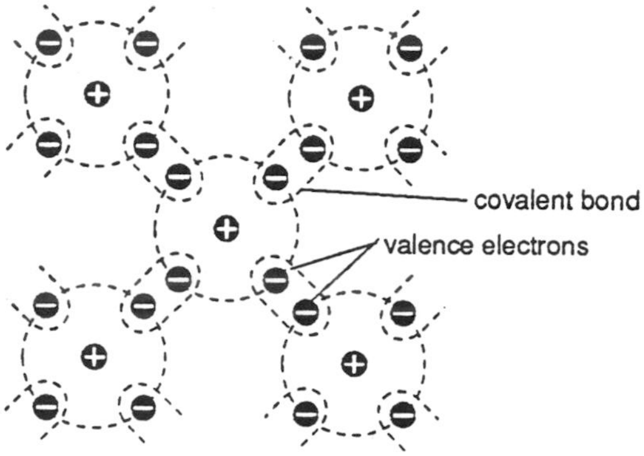

covalent bond

valence electrons

(b) p-type material

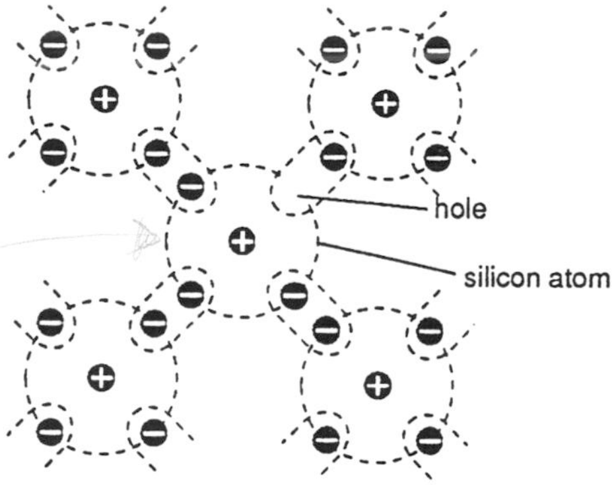

hole

silicon atom

(c) n-type material

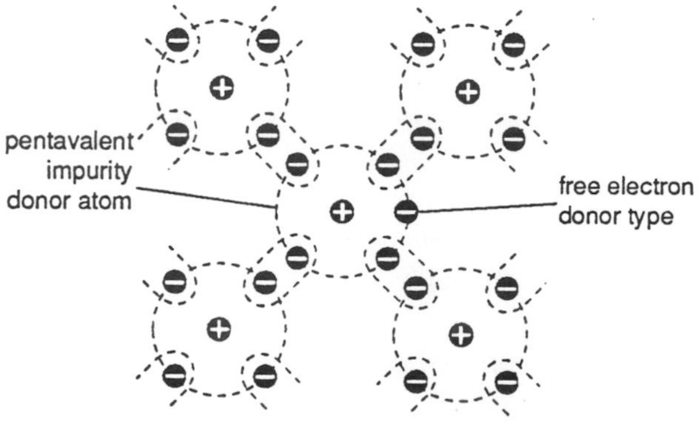

pentavalent impurity donor atom

free electron donor type

- In materials that are good insulators the electrons are more firmly attached and cannot be forced away from the nucleus.
- An atom that has lost electrons becomes positively charged and seeks to attract negative electrons.
- Electrons can be loosened and a significant flow of electrons can be produced by thermal, magnetic or chemical means.
- Semiconductor materials are perfect insulators at absolute zero temperature; however, at room temperature electrons break loose due to thermal agitation.
- The conductivity of semiconductor materials can be improved by doping. The resulting material is used in the production of semiconductor devices such as transistors, thyristors, etc.

Test 1

Choose which of the four answers is the correct one.

(1) When the number of protons in the nucleus is equal to the number of orbiting electrons the atom is said to be:

(a) positively charged,
(b) negatively charged,
(c) in electrical balance,
(d) at equipotential.

(2) Any atom that has lost electrons becomes:

(a) positively charged,
(b) negatively charged,
(c) in electrical balance,
(d) at equipotential.

(3) A form of energy that does not produce an electron flow is:

(a) thermal energy,
(b) magnetic energy,
(c) chemical energy,
(d) kinetic energy.

(4) The most common form of semiconductor material in use is:

(a) arsenic,
(b) silicon,
(c) antimony,
(d) gallium.

(5) A pentavalent impurity substance used for doping semiconductor material is called:

(a) arsenic,
(b) silicon,
(c) indium,
(d) gallium.

(6) A trivalent impurity substance used for doping semiconductor material is called:

(a) arsenic,
(b) silicon,
(c) antimony,
(d) gallium.

(7) An atom is composed of:

(a) a nucleus and one or more elements,
(b) ions and electrons,
(c) a nucleus and one or more electrons,
(d) protons and neutrons.

(8) A nucleus is composed of:

(a) a nucleus and one or more elements,
(b) ions and electrons,
(c) a nucleus and one or more electrons,
(d) protons and neutrons.

(9) The charge of electricity carried by a proton is:

(a) positive,
(b) negative,
(c) high,
(d) low.

(10) The known elements number:

(a) 920,
(b) 92,
(c) 52,
(d) 32.

Chapter 2
Passive Electronic Components

2.1 Electronic component recognition

British Standard (BS) 3939 gives recommended graphical symbols for identification of electronic components in much the same way as it does for electrical accessories in electrical circuits. Information Sheet No. 2A shows examples of the more common ones of these. Symbols used in circuit diagrams for foreign products may vary a little from these.

It is essential when replacing faulty electronic components in a circuit, that they are replaced with identical components, or the exact equivalent components having the same characteristics as the original. It is therefore essential that any electrician contemplating the repair of an electronic device knows something about the various components and is able to recognize them and identify the values of them.

Manufacturers use various methods of identifying components and we shall discuss these as we deal with the particular components.

2.2 Resistors (fixed value)

In electrical engineering, resistance in a circuit is a major disadvantage; in cables, across contacts, battery terminals, etc. There are numerous regulations to be followed to avoid excessive voltage drop and power loss in cables.

In electronics, however, resistance is a distinct advantage; in fact, electronics would not exist in its present form if it were not for resistance and resistors.

A resistor is an electronic component that is specially manufactured from materials chosen for their resistive properties. They are used in electronic circuits for a number of purposes such as voltage dividing, changing and varying voltages and current limiting to name just a few. The most common types of resistors are listed below:

- Carbon compound;
- Carbon film;
- Metal oxide;
- Metal film;
- Cermet;
- Wirewound.

Examples of these are shown on Information Sheet No. 2B.

Information sheet No.2.A.

Resistor symbols

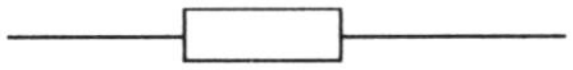

resistor general symbol

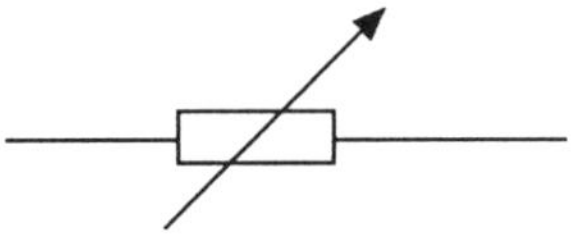

variable resistor

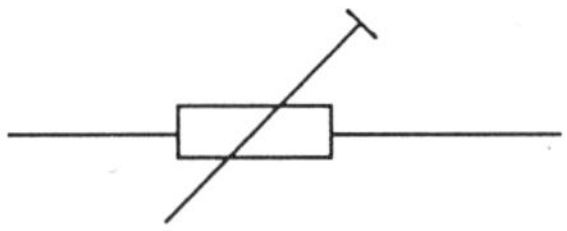

preset resistor

Manufacture of resistors

Carbon compound resistors are made by mixing together fine grains of carbon with an insulating filler. The mixture is then formed into rods and heated in a kiln. The ratio of impurity to carbon dictates the final value of resistance. Silver coated end caps with tinned copper wires attached are fitted to the ends of the rod to facilitate connection to the circuit. The resistor is then enclosed in an insulated tube of plastic or ceramic and given a coat of paint to protect it from the damp. Carbon resistors have power ratings of between 0.1 and 2.0 W. This should not be exceeded as overloading will cause heating, and, since carbon has a negative temperature coefficient, the value of the resistor will drop, resulting in breakdown of the component and possibly other components too. Because the carbon resistor is not very stable and can vary with temperature and old age, other methods of producing resistors have evolved.

Carbon film, metal oxide and metal film resistors, although different in their characteristics, are all made in much the same way. An even film of resistive material such as pure carbon (carbon film), nickel chromium (metal film) or metal and insulating oxide (metal oxide) is deposited onto a ceramic rod. The quality of the ceramic rod must be high because otherwise it may degrade the property of the finished resistor. The materials used must have similar expansion properties to prevent cracking. A spiral is then cut along the deposited film to give the final value of resistance. This is a very accurate way of manufacturing resistors and tolerances of 1.0% or better are common.

Cermet resistors are a mixture of ceramics and a metal contained in an insulating cylinder. The ratio of ceramic to metal governs the final value of the resistor. This type of resistor is capable of power dissipation of 2.0 W or greater.

When it is necessary for a resistor to dissipate really large amounts of power then *wirewound types* should be used. These are made by winding a resistance wire of known value onto an insulating former and encapsulating it in a coating of vitreous enamel. The alloy resistance wire is usually made of nickel chromium (Nichrome) or copper-nickel (Eureka) which results in power ratings of up to 20 W. This type of resistor runs at high temperatures and care should be taken that this does not affect surrounding components. Heat reduction is often achieved by fitting the resistor in an aluminium case and bolting it down to a heat sink.

Circuit symbols for resistors (fixed value) can be found on Information Sheet No. 2A.

2.3 Standard resistor colour-coding

The fixed value resistors have their values marked upon them in the form of colour bands (see Information Sheet No. 2C). The marking of the resistor in this way ensures that its value can be seen whichever way it is mounted on the circuit board. The markings are at one end of the resistor and if the resistor is turned so that these are on the left, then the colour bands can be read from left to right.

Information sheet No.2.B.

Typical resistor types

Carbon resistor

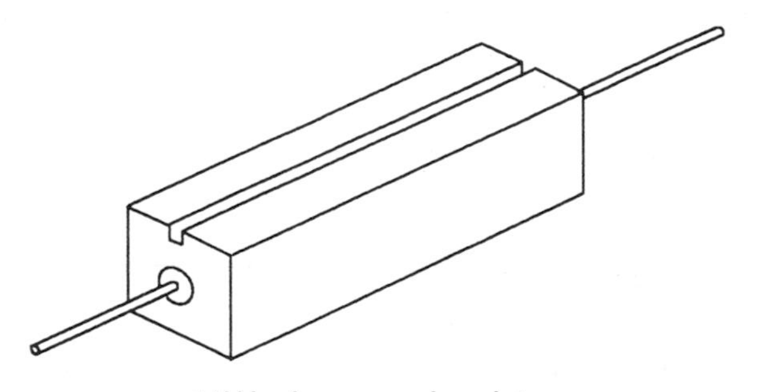

11W wire wound resistor

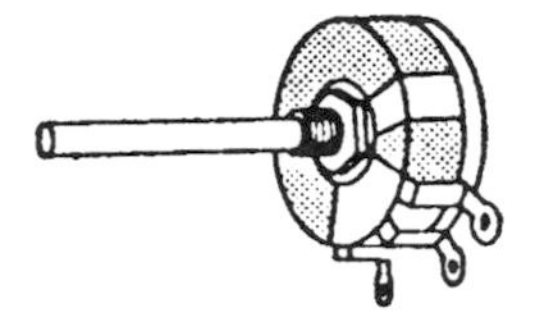

Variable resistor

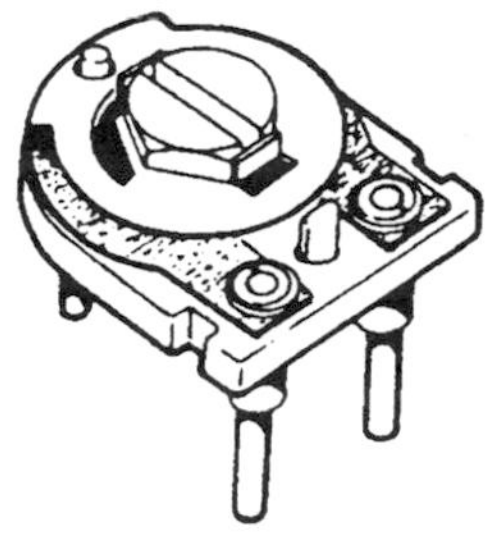

Preset resistor

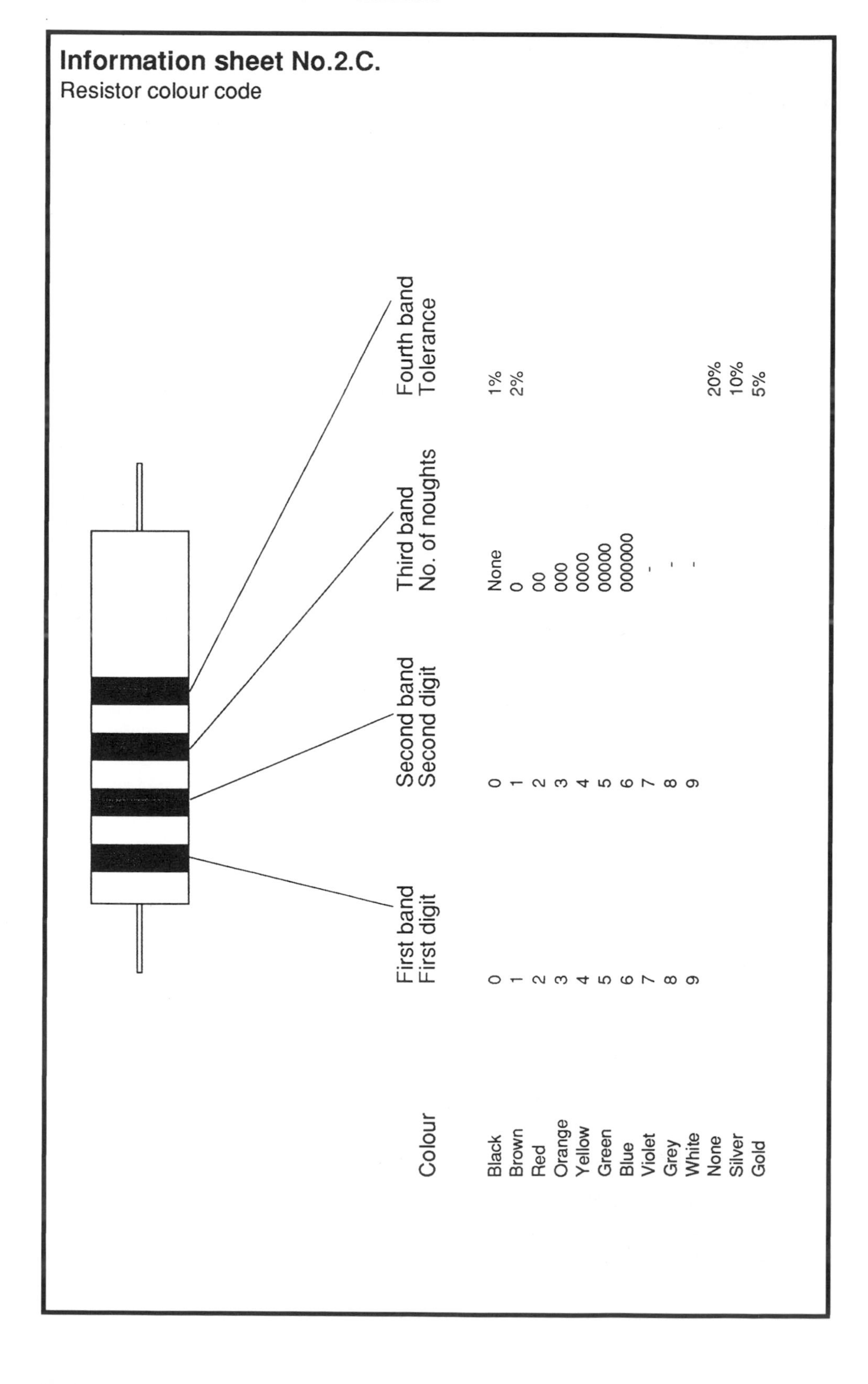

Information sheet No.2.C.
Resistor colour code

Colour	First band First digit	Second band Second digit	Third band No. of noughts	Fourth band Tolerance
Black	0	0	None	1%
Brown	1	1	0	2%
Red	2	2	00	
Orange	3	3	000	
Yellow	4	4	0000	
Green	5	5	00000	
Blue	6	6	000000	
Violet	7	7	-	
Grey	8	8	-	
White	9	9	-	
None				20%
Silver				10%
Gold				5%

We can see from Information Sheet No. 2C that:

- The first colour band indicates the first digit of the resistance value.
- The second colour band indicates the second digit of the resistance value.
- The third band (sometimes called the multiplier) indicates the number of noughts to be added after the first two numbers.
- The fourth colour band (which has a gap between it and the other bands) indicates the resistor tolerance.

Tolerance

Due to inaccuracies during manufacture it is impossible to produce resistors to the precise value indicated by the colour bands; they can only be produced to within certain tolerances. These are marked by coloured bands as indicated on Information Sheet No. 2C, common ones being gold or silver which are 5% or 10% respectively. Good quality resistors used for accurate work have tolerances of 1% or 2% and are marked **brown** or **red** respectively. If no fourth colour band is present then the tolerance is assumed to be 20%. It is important to remember that the final value of the resistor can be plus or minus the tolerance value, so that a resistor of 100 Ω with a tolerance of 10% could in fact have a value of between 90 and 110 Ω.

Using the standard colour code

EXAMPLE 2.1

A resistor has the following colour bands:

First band	Red
Second band	Violet
Third band	Orange
Fourth band	Gold

What is the value of the resistor and its tolerance?

From Information Sheet No. 2C:

Red	= 2	The first digit is then	2
Violet	= 7	The second digit is then	7
Orange	= 3	The number of noughts is	3

This gives a value of 27 000 Ω.

The fourth band is **gold**; from Information Sheet No. 2C **gold** is 5% so the value is 27 000 Ω ± 5%. This is written as:

27 kΩ @ 5%

EXAMPLE 2.2

A resistor has **brown, black** and **red** colour bands. What is its value and tolerance?

Brown = 1
Black = 0
Red = 00

This gives a value of 1000 Ω. The fourth band is absent so a tolerance of 20% is assumed. The value is therefore:

1 kΩ @ 20%

EXAMPLE 2.3

A resistor has **yellow, violet, orange** and **gold** colour bands. What range of values may the resistor have?

According to Information Sheet No. 2C the value is 47 kΩ @ 5%

Now 5% of 47 000 is 2350. Since the resistor value can be plus or minus this, then the value may be between:

47 000 + 2350 = 49 350 or,

47 000 − 2350 = 44 650

If we were to measure the resistor using an ohmmeter, then the reading would fall somewhere between 49.35 and 44.65 kΩ. Any reading within this range would meet the manufacturer's specifications.

EXAMPLE 2.4

A resistor has **brown, grey, brown** and **silver** colour bands.

If we were to measure its resistance with an ohmmeter, what range of values would you expect it to fall within?

The nominal value, according to Information Sheet No. 2C is 180.

The fourth band is **silver** which means a tolerance of 10%. Now 10% of 180 is 18. Since the resistor value can be plus or minus this, then the value can be between:

180 + 18 = 198 Ω or,
180 − 18 = 162 Ω

EXAMPLE 2.5

A good quality resistor has a tolerance of 2% and a nominal value of 470 Ω. What would be the colour coding for such a resistor?

4 = Yellow
7 = Violet

0 = Brown (one nought)
2% = Red

Therefore, the colour coding would be, from left to right: **yellow**, **violet**, **brown** and **red**.

It is important that the reader remembers the colours in order, and the value each represents. Perhaps the best way is to use this well-known rhyme as a memory aid:

Bye Bye Rosie, Off You Go, Bristol Via Great Western.

The rhyme does not mean much until you realize that the first letter of each word is the first letter of each colour in the correct order. The tolerance colours you will just have to remember, unless you make your own rhyme up for them.

Remember! Tolerance is a percentage of the marked value.

2.4 BS 1852 letter code for resistors

This method of indicating the values of resistors on circuit diagrams was evolved in order to avoid confusion with faintly marked or missing decimal points. For example, if a resistor should be 4.7 kΩ and, due to the decimal point being faint or left out, the reader assumed it to be 47 kΩ, this would clearly be wrong and could have disastrous results.

In the BS 1852 method, where the value of the resistor includes a decimal point, this is replaced by the prefix for the multiplication factor.

For all resistors with values below 1 kΩ, R is used, so that:

0.47 Ω becomes R47
1 Ω becomes 1R0
8.2 Ω becomes 8R2
10.2 Ω becomes 10R2
820 Ω becomes 820R

For all resistors with values in kilohms, K is used so that:

1 kΩ becomes 1K0
8.2 kΩ becomes 8K2
10.2 kΩ becomes 10K2
820 kΩ becomes 820K

For all resistors with values in Megohms, M is used so that:

1 M Ω becomes 1M
8.2 M Ω becomes 8M2
and so on.

Additionally, a letter may be added to the end of the code to indicate the tolerance, the letters being:

F = 1%
G = 2%
J = 5%
K = 10%
M = 20%

For example, 5.6 kΩ @ 5% becomes 5K6J and 10 kΩ @ 2% becomes 10KG.

2.5 Preferred values for resistors

In the early days of electronics resistors were produced in standard values of 10 Ω, 25 Ω and 50 Ω with their multiples of ten. This resulted in the production of far more resistance values than were really necessary. This was due to the fact that although the old resistors were marked with their nominal values, there was a certain amount of overlap due to tolerance. Manufacturers soon realized that they could cover all the values required with fewer resistors by carefully spacing out the values and taking into account the tolerances. We have already seen that a resistor of value 100 Ω at a tolerance of 10% could have a value between 90 and 110 Ω. The next sized resistor in this tolerance range that we could use without too much overlap would be 120 Ω @ 10%, giving a range of 108 and 132 Ω. Resistors are now manufactured to carefully worked out values which are known as *preferred values* and Information Sheet No. 2D gives full details of these.

2.6 Variable resistors

Variable resistors are very useful devices as their value can be varied in accordance with the requirements of the circuit. They consist of a track of resistance material and a movable wiper which either increases the amount of resistance material in circuit or alternatively reduces the amount by cutting some of it out. The wiper arm is adjusted by either a shaft, as in the rotary types used for volume controls, etc., or a screwdriver slot, as in the preset types or potentiometers (pots) used to adjust the potential difference in a circuit.

Three common materials used for the production of variable resistances are:

- Carbon,
- Wirewound,
- Cermet.

Information sheet No.2.D.

Preferred values

E24 series

10 11 12 13 15 16 18 20 22 24 27 30 33 36 39 43 47 51 56 62 68 75 82 91

E12 series

10 12 15 18 22 27 33 39 47 56 68

E6 series

10 15 22 33 47 68

A variable resistor may be connected in one of two configurations:

(1) to vary the current in a circuit (when it is called a rheostat)
(2) to vary the voltage in a circuit (when it is called a potentiometer).

Circuit symbols for variable resistors are to be found on Information Sheet No. 2A.

2.7 Non-linear resistors

Linear resistors, sometimes known as ohmic resistors, are ones that obey Ohm's Law. This means that if we increase the voltage then the current will increase too, in accordance with the law. Metallic conductors and carbon obey Ohm's Law if their temperature is kept fairly constant. Lamp filaments, cells, electrolytes and semiconductors do not obey this law. Three types of non-linear resistor are met in electronics:

(1) *Voltage dependent resistors (VDRs)* as their name suggests, have a resistance value that depends upon the potential difference across them. They are made by pressing silicon carbide with a ceramic binder and are used as surge limiters and voltage stabilizers.
(2) *Light dependent resistors (LDRs)* are also called cadmium sulphide (CdS) cells. In fact, they are not really cells because they are not a source of electromotive force (emf). The resistance of the LDR falls considerably when exposed to light. The same effect is produced with radiation just outside the visible spectrum (infrared and ultraviolet). This device has many light-actuated switching applications.
(3) *Thermistors*, whose name is derived from the words *thermal* and *resistor*, have a resistance which varies with changes in temperature. If the resistance increases with temperature rise then a thermistor is said to have a positive temperature coefficient (PTC). If the resistance decreases with temperature rise then it is said to have a negative temperature coefficient (NTC). Thermistors of the NTC type seem to be the more common. They are made from oxides of manganese and nickel, with small amounts of iron, copper, and cobalt to vary the properties as required.

Large thermistors are used to prevent *surges* when certain electrical equipment is switched on. Small thermistors can be used for sensitive temperature measurement and control.

Circuit symbols for non-linear resistors can be found on Information Sheet No. 2E.

Information sheet No. 2.E.
Non-linear resistors

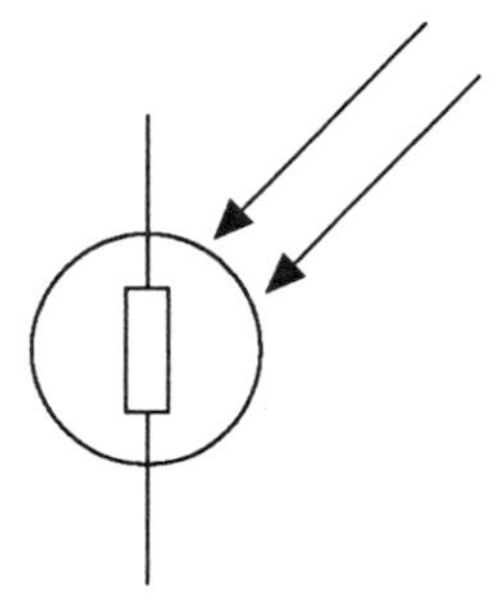

Light dependant resistor

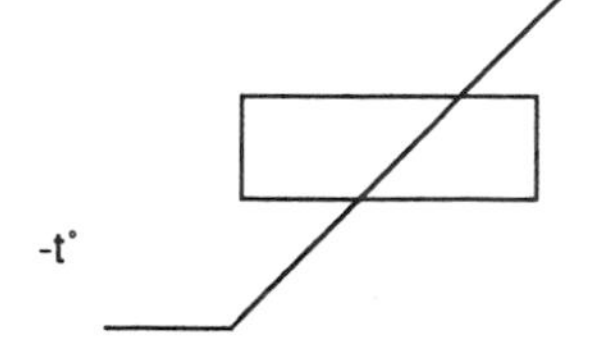

Negative temperature coefficient resistor

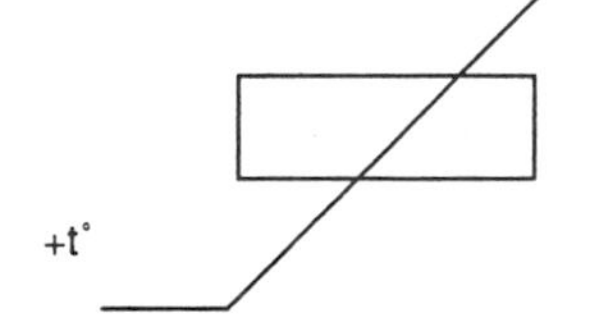

Positive temperature coefficient resistor

SUMMARY

Sections 2.2–2.7 have described in simple terms resistors used in electronic circuits. The following important points should have been learned:

- That electricians should be able to recognize and identify the different types of resistors;
- How to use resistors in electronic circuits;
- How to determine the value of resistors using the standard colour code;
- An understanding of the term *tolerance* and how to carry out calculations involving this;
- How to determine the value of resistors using the BS 1852 letter code;
- An understanding of the expression *preferred values*;
- How to recognize BS 3939 symbols used for the different types of resistors on circuit diagrams.

2.8 Capacitors

Take a look at any electronic circuit diagram and capacitors seem to appear in large numbers. In simple terms a capacitor consists of two metal plates that are separated by insulating material. The insulating material is given the special name of dielectric.

If an uncharged capacitor is connected across a battery (or other voltage source) the emf of the source will cause electrons to drift from one plate to the other until the potential difference (PD) between the plates is equal to the PD of the source. The capacitor at this point is said to be charged (see Information Sheet No. 2F).

When the discharged capacitor is initially connected to the source of supply, a large current (I) flows for an instant, and then reduces in t seconds as the capacitor becomes fully charged (Q). The current in the ideal capacitor will have fallen to zero at this point. The amount of charge transferred can be determined from:

$$Q = It \text{ (coulombs)}$$

where Q is the charge in coulombs, I is the capacitor current, and t is the time(s).

When the capacitor is disconnected from the source of supply it will remain charged until the charge leaks away. This could take a long time, hence the danger of touching charged capacitors in recently operated high voltage equipment. If a discharge resistor is connected across the terminals of a capacitor the electrons will flow in the opposite direction until the PD across the plates has fallen to zero.

The ability of the component to store an electric charge is called its *capacitance*.

Information sheet No.2.F.
Charge on a capacitor

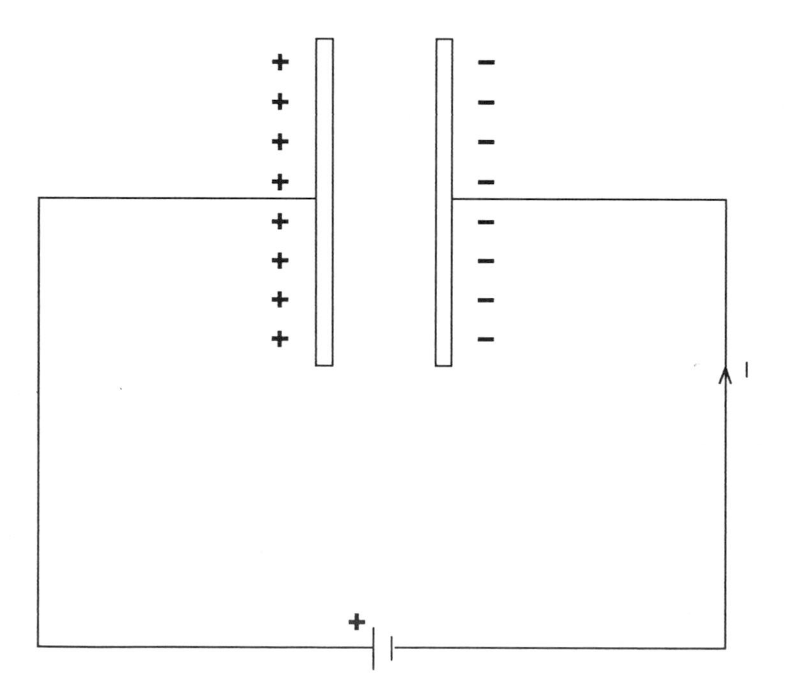

Electron flow is in the direction shown by the arrow. Conventional current flow is in the opposite direction.

By definition a capacitor has a capacitance of one farad when a PD of 1 V maintains a charge of one coulomb (C) on that capacitor, so that:

$$C = \frac{Q}{V} \text{ (farads)}$$

or

$$Q = CV$$

where Q is the charge in coulombs, C is the capacitor value in farads, and V is the voltage across the capacitor.

Units of capacitance

It will be seen that the value of capacitors is expressed in units called farads (F); these are named after the British scientist Michael Faraday. The number of farads is an indication of how much electrical energy can be stored in the capacitor. The larger the value of the capacitor, the more energy it can store. The farad is a rather large unit and so it is more usual in electronics to rate capacitors in fractions of a farad, so that:

one millionth	= a microfarad	= 1 μF	=	1×10^{-6} F
one thousand millionth	= a nanofarad	= 1 nF	=	1×10^{-9} F
one million millionth	= a picofarad	= 1 pF	=	1×10^{-12} F

2.9 Types of capacitor

Paper In these capacitors the dielectric material between aluminium foil electrodes consists of waxed paper. This is usually rolled up like a swiss roll and placed in an aluminium can. Paper capacitors are easily recognized and their values, which range from 0.001 μF to 8 μF, are marked on the can. They are suitable for medium frequency work, particularly mains interference suppression and power factor correction.

Plastic film These capacitors can use one of four dielectrics:

(1) polypropylene,
(2) polyester,
(3) polycarbonate,
(4) polystyrene.

With the exception of polystyrene, which has a low melting point, these plastics have a metallic film deposited on them by a vacuum evaporation process. This is what is meant by the term metalized capacitors. The range of values depends on the type of construction. Typically, a polycarbonate film construction hermetically sealed in a hot tinned brass case with epoxy resin and insulated with a polyester sleeve can range from 1 μF to 10 μF. Capacitors of this type are suitable for professional and military purposes and find use in timing and other high stability applications.

Silvered mica These rather expensive capacitors are made by depositing a thin layer of silver on either side of a thin sheet of mica. Their normal range of values is 1 pF to 0.01 μF and they are very suitable for high frequency work, tuned circuits, filters, etc. They are very stable and are accurate to within 1% of their marked value.

Ceramic These capacitors consist of a silver-plated ceramic plate or disc. They can be made to give a relatively large capacitance in a small space and are suitable for high frequency work. They come in a number of different forms. Typically a multilayer ceramic chip, surface mounting capacitor, with metalized silver terminations and tin coated nickel barrier to give increased solderability, will have a range between 10 pF to 0.22 μF.

Electrolytic Where large values of capacitance are required then electrolytic capacitors are used. These devices are available in the range: 0.47 μF - 47000 μF and greater. These capacitors are polarized, which means they have a positive end and a negative end and must be connected in circuit the correct way round.

Two main types of material - aluminium and tantalum - are commonly used in the construction of electrolytic capacitors:

- *Aluminium.* These capacitors are made by immersing a foil of very pure aluminium in an electrolyte bath. A fixed voltage is applied and the flow of current builds up a layer of aluminium oxide on the surface of the aluminium. The layer of aluminium oxide, which is about eight times thinner than human hair, acts as a dielectric. It is then wound with tissue and another layer of foil to form the capacitor.
- *Tantalum.* These capacitors are much smaller than the corresponding aluminium type, consisting of an oxide layer on tantalum foil electrodes. Tantalum electrolytic capacitors are also polarized, and have high capacitance and low leakage current. They are usually made in bead form and are colour-coded for easy identification. A selection of the different forms of capacitors is shown on Information Sheet No. 2G. The circuit symbols for capacitors are shown on Information Sheet No. 2H.

2.10 Working voltage for capacitors

Breakdown voltage is the voltage at which the dielectric of the capacitor will break down. Capacitors are marked with a *working voltage*, e.g. 100 V wkg, which is the

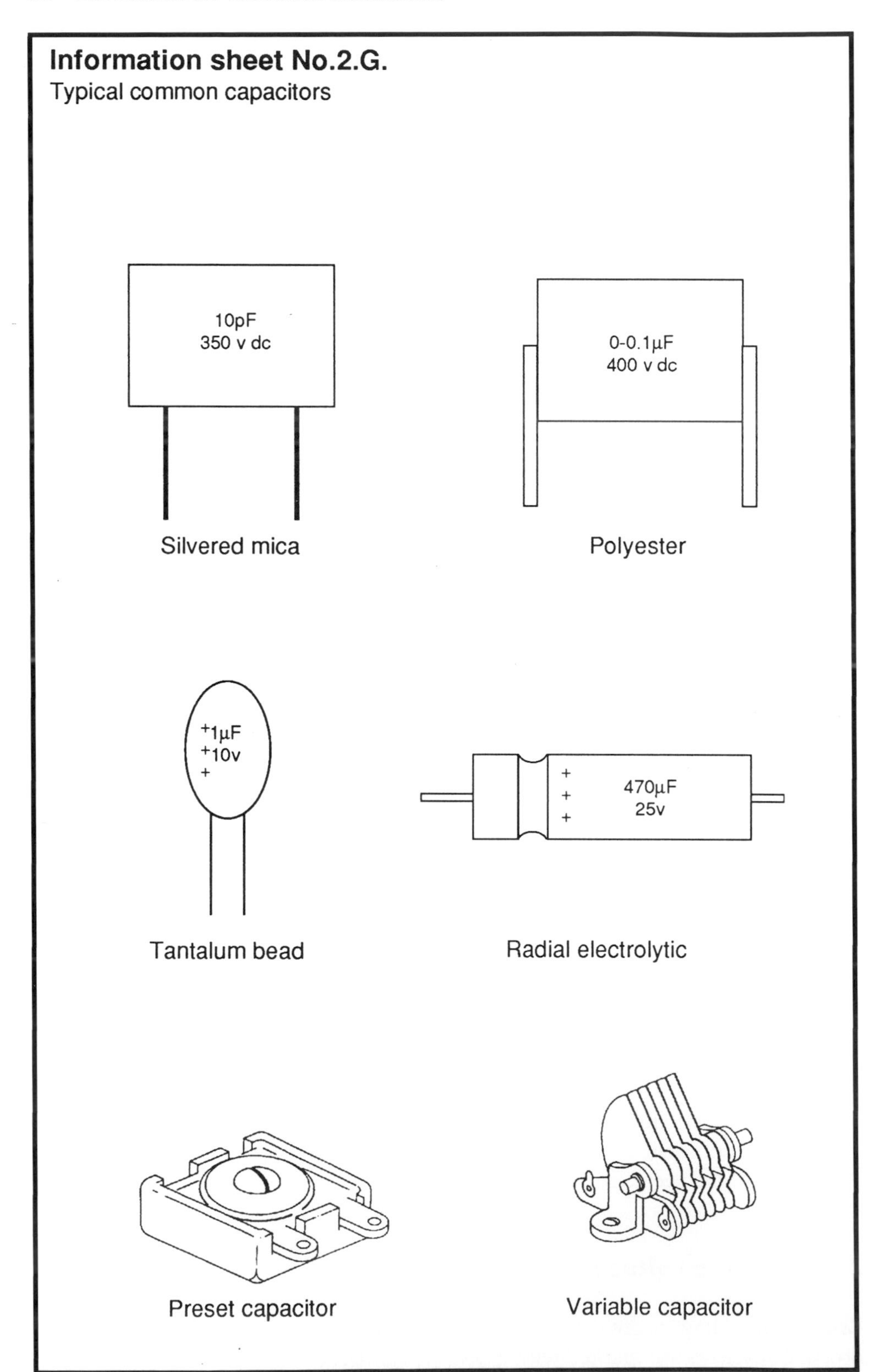
Information sheet No.2.G.
Typical common capacitors
10pF
350 v dc
Silvered mica
0-0.1μF
400 v dc
Polyester
+1μF
+10v
+
Tantalum bead
+
+
+
470μF
25v
Radial electrolytic
Preset capacitor
Variable capacitor

Information sheet No.2.H.
Capacitor symbols

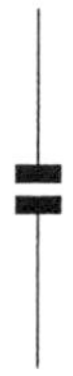

capacitor general symbol

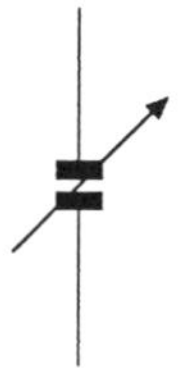

variable capacitor

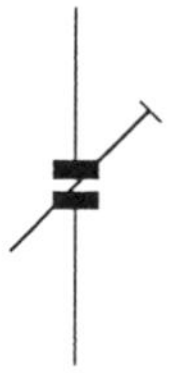

preset capacitor

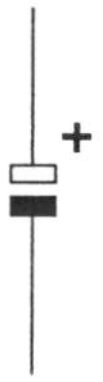

electrolytic capacitor

non polarised electrolytic capacitor

maximum voltage that the capacitor must be subjected to. It is important to remember that, when connecting capacitors to alternating current (a.c.) supplies, these are usually expressed as the root mean square (rms) values. In fact, at 240 a.c. the voltage is approximately 340 V maximum, so a capacitor with a 400 V working voltage should be chosen.

Here are some important points to bear in mind regarding capacitors in use:

- Electrolytic capacitors are polarized and must be connected the correct way round as shown on the circuit diagram.
- If connected wrongly, electrolytic capacitors can explode; if this occurred it would be due to gases being formed as the oxide was removed from the foil.
- Manufacturers clearly mark at least one end of the electrolytic capacitor to assist with connection.
- Capacitors are available with either axial connections (the leads stick out either end) or radial (both leads are at one end).
- The tolerance of aluminium type electrolytic capacitors is quite large – possibly –20% to +50% – as, in practice, high upper tolerance is not important.
- Capacitors have maximum working voltages and as a rule of thumb this should be about twice the operating voltage.
- The maximum working voltage of electrolytic capacitors is unlikely to be above 100 V. If this is exceeded they can explode.

2.11 Colour-coding of capacitors

Capacitors usually have their value marked on the side of the component, along with the maximum working voltage. However, some polyester capacitors use colour codes to show their value, tolerance and maximum working voltage. The colour coding is similar to that used for resistors, but a fifth band is used to indicate the maximum working voltage. This code is shown on Information Sheet No. 2I and is read from the top of the capacitor to the bottom.

Using the colour code

EXAMPLE 2.6

What is the value, tolerance and working voltage of a polyester capacitor colour-coded **white, blue, red, black, yellow**?

From the table on Information Sheet No. 2I:

White = 9
Blue = 6
Red = 00

Information sheet No.2.I.
Capacitor colour code

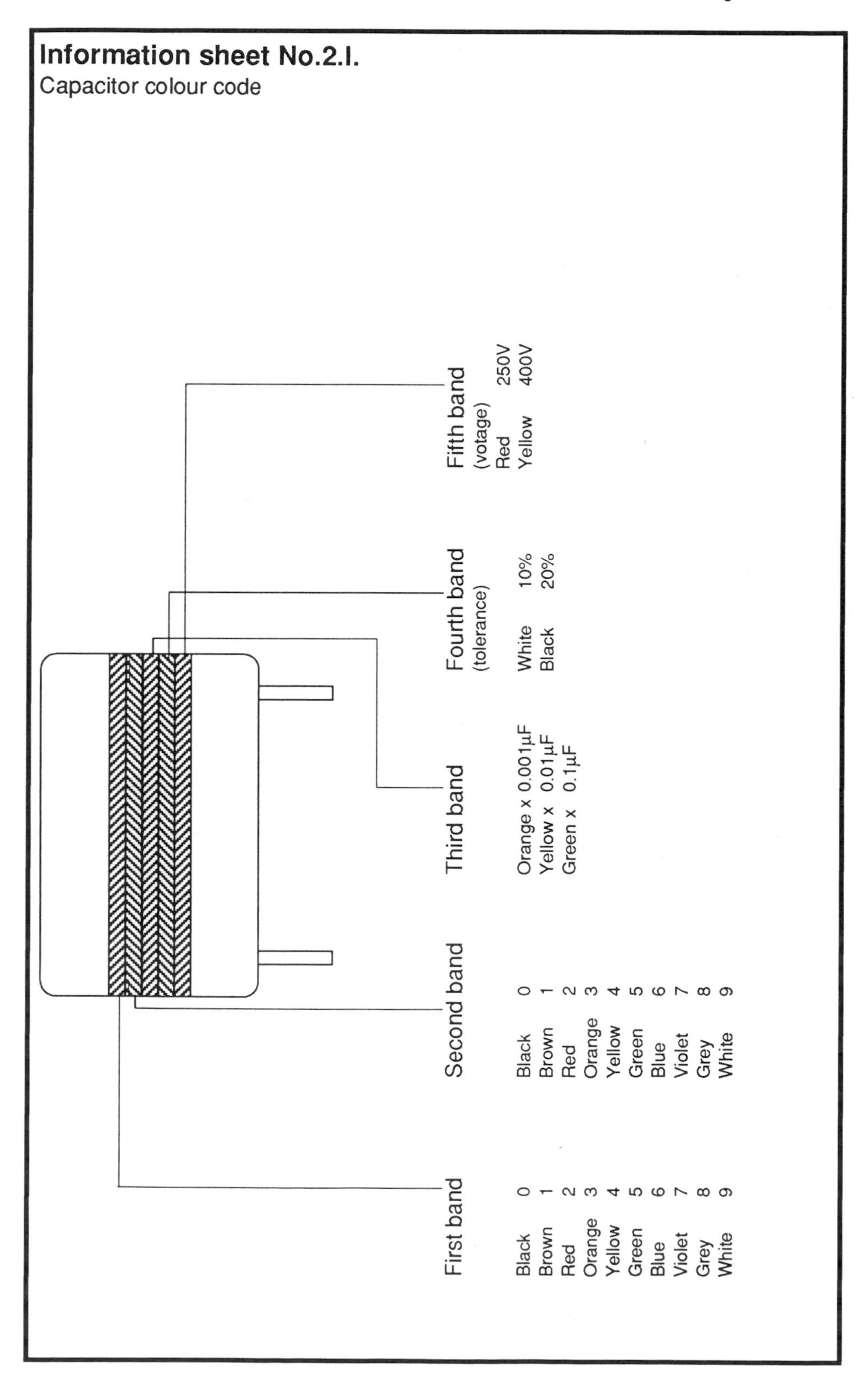

Black = 20%
Yellow = 400 V d.c.

This gives a value of 9600 pF, at a tolerance of 20%, with a working voltage of 400 V d.c.

EXAMPLE 2.7

A polyester capacitor has a value of 100 000 pF, a tolerance of 20% and a maximum working voltage of 250 V d.c. What will be the colour-coding for this capacitor?

First band = brown
Second band = black
Third band = yellow
Fourth band = black
Fifth band = red

Tantalum capacitors use a similar colour code, a coloured spot being used to identify the number of noughts to be added.

2.12 Capacitor letter code

The letter code used in conjunction with capacitors is generally as described for resistors (see Section 2.4) with certain additions. For capacitors the position of the decimal point and the value of the decimal multiplier is given by the following alphabetical characters:

m = one thousandth = 10^{-3}
u = one millionth = 10^{-6}
n = one thousand millionth = 10^{-9}
p = one million millionth = 10^{-12}

Thus a value of 1 pF is written 1 p0, and 10 nF is written 10 n.

2.13 Variable capacitors

Two types of variable capacitor exist: the rotary type and the compression type.

The rotary type consists of two sets of plates situated very close together but separated by air which, being an insulator when dry, forms the dielectric. One set of plates is fixed; the other set is mounted on a shaft which is movable. When the shaft is rotated the plates move and the area between the two sets of plates alters. In this way the capacitance of the device is altered.

The capacitance of the compression type is varied by altering the distance between the plates. This is usually achieved by the use of a screwdriver in a slot provided.

Capacitors behave differently when connected to a.c. supplies than when connected to d.c. supplies. A.C. circuits will be discussed in Chapter 7.

SUMMARY

Sections 2.8–2.13 have described in simple terms capacitors used in electronic circuits. The following important points should have been learned:

- That electricians should be able to recognize and identify the different types of capacitors;
- How capacitors are used in electronic circuits;
- How to determine the value of capacitors using the colour code;
- How to recognize BS 3939 symbols used for the different types of capacitor on circuit diagrams;
- How to determine the value of resistors using the BS 1852 letter code;
- An understanding of the expression 'working voltage' and its implication for a.c. circuits;
- The advantages and disadvantages of the different types of capacitor.

2.14 Inductors

Not surprisingly, inductors have the property of inductance. This is the property of a circuit which tends to prevent any change of current in that circuit.

In its simplest form an inductor consists of a length of insulated wire wound around a former, which can be a plastic rod or tube. The inductor can have a core inside the tube, or just have air as a core.

When an inductor is connected to a direct current (d.c.) supply, a current flows through the winding causing a magnetic field to grow around the coil. The effect of this is to induce current into itself. The self-induced current will be opposite in polarity to the supply current, and so will tend to oppose the current from the supply. When the supply is first connected the coil appears to have a high resistance to current flow. Once the magnetic field is established, its resistance to current falls. The current through an inductor is shown in Fig. 2.1.

The field created is a magnetic field and so electromagnetic energy will be stored in it. If the inductor is large then considerable amounts of energy will be stored in the field and care should be taken even when the circuit is switched off.

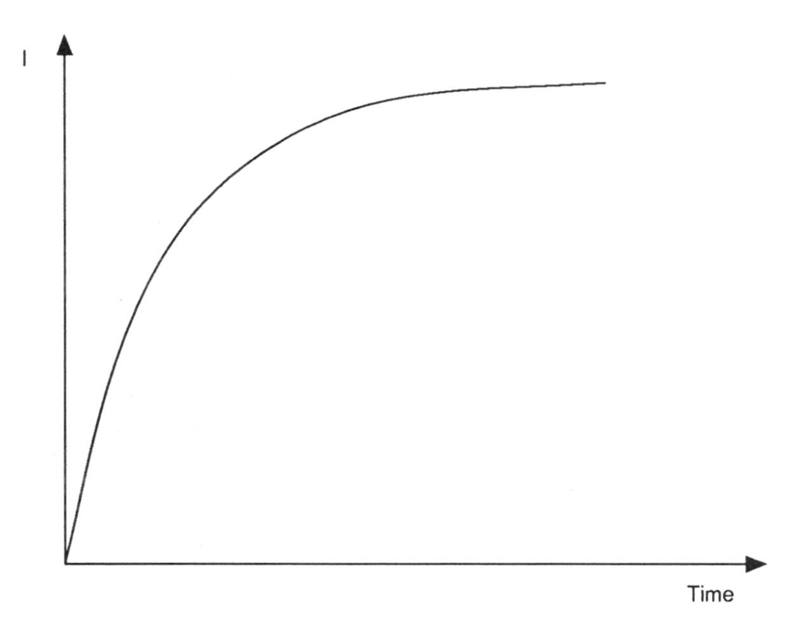

Fig. 2.1 Current through an inductor at switch on.

The magnitude of the inductance is dependent upon two main factors:

(1) The number of turns in the coil of the device;
(2) The material, if any, that the core is made from.

The inductance may be increased many times by the inclusion of a core of some magnetic material (such as iron), although high frequency inductors are usually air cored.

Because the inductor is made from many turns of wire, it not only possesses inductance but also resistance. The resistance will depend on the cross-sectional area of the wire and the number of turns on the coil. Not many inductors are in use today in electronic circuits, mainly because of their high current demand. The unit of inductance is the henry (H).

2.15 Types of inductor

Low current fixed inductors These are general-purpose inductors suitable for power decoupling in logic circuits, filters, etc. They are constructed from fine polyurethane-insulated copper wire wound onto a ferrite bobbin. They have an inductance range of 100–1000 μH @ 10%.

High current fixed inductors These are designed for use in power line applications such as switching regulators, filters and suppression circuits. They are constructed from a special high-saturation flux density material enabling them to be used in high-current applications. The coil is terminated with pretinned, enamelled copper leads to facilitate connection to chassis or PCB. They have an inductance range of 100–470 μH @ 10%.

Ferrite cored inductors This type come in a number of different forms. One type consists of a ferrite-based coil former encapsulated in a polypropylene outer case. Another consists of a coil mounted on a polyphenylene oxide moulded former and containing a ferrite core. They can have inductance ranges of 1–1000 μH.

Large inductors These are constructed much like a transformer. Here the coil is wound onto a special shaped core made up of lots of iron plates or stampings and sandwiched together, yet insulated from each other. The plates are called laminations and make the inductor more efficient. The choke of a fluorescent fitting is a form of inductor.

2.16 Variable inductors

Two main types exist:

(1) An inductor in which the length of the ferrite core can be changed for one of smaller or greater length
(2) An inductor in which the ferrite core is threaded and can be screwed in or out of the former (see Fig. 2.2).

In either case the longer the ferrite rod inside the core the higher will be the inductance. Circuit diagram symbols for inductors are shown on Information Sheet No. 2J.

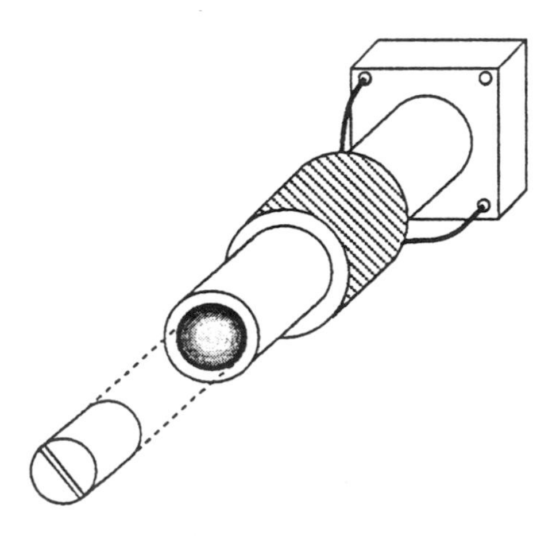

Fig. 2.2 Inductor with ferrite core.

Information sheet No.2.J.

Inductor symbols

inductor general symbol

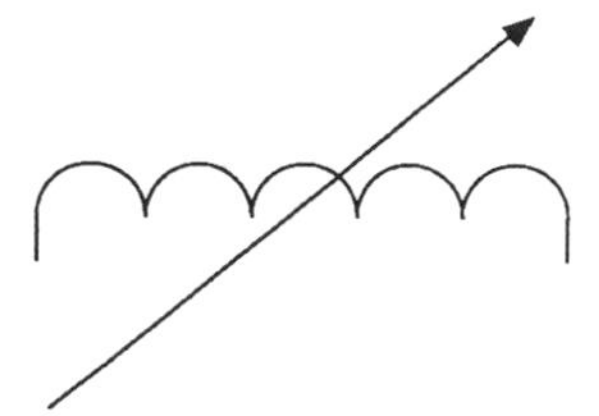

variable inductor

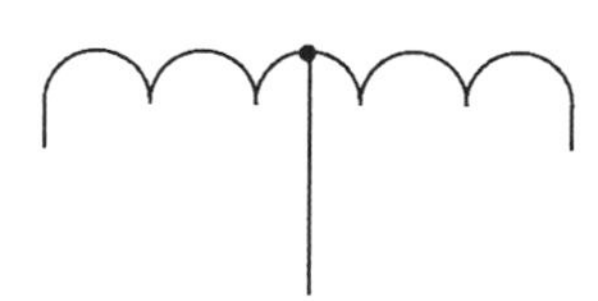

inductor with tapping

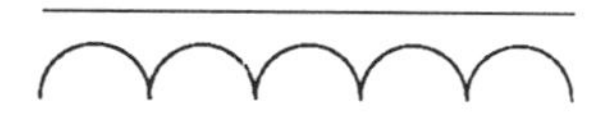

inductor with iron core

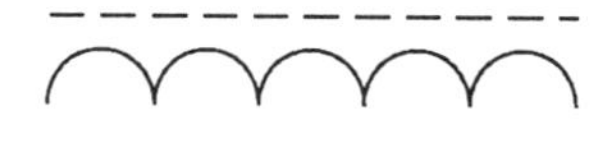

inductor with dust core

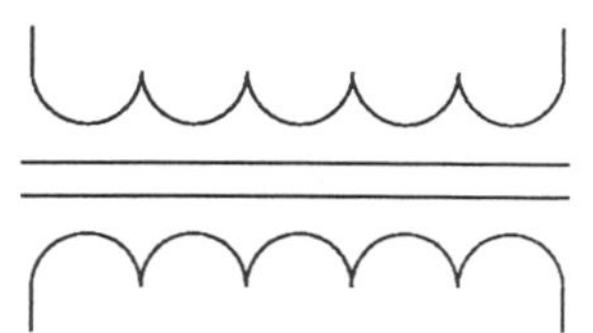

transformer with iron core

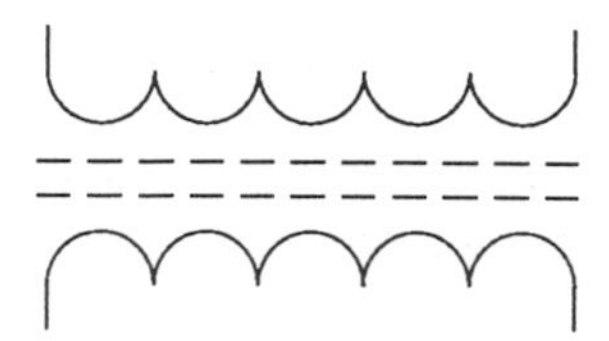

transformer with dust core

Inductors behave differently when connected in a.c. circuits than when connected in d.c. circuits. A.C. circuits are discussed in Chapter 7.

2.17 Transformers

A transformer is a device which, by electromagnetic induction, transforms alternating or intermittent power from one voltage to another. Students will be familiar with the use of transformers at higher voltages and power levels; however, in electronics we only concern ourselves with transformers having power ratings of a few volt-amps (VA).

In simple terms, a transformer consists of a primary winding on the input side of the transformer, a secondary winding on the output side of the transformer and a core. The core can be of air, iron or ferrite depending upon the application. The ratio of input voltage to output voltage depends on the ratio of the turns on the primary winding to the turns on the secondary winding (neglecting losses). If there are fewer turns on the secondary winding than on the primary winding, then the output voltage will be smaller. If the situation is reversed then the voltage will be greater.

EXAMPLE 2.8

A transformer has 1500 turns on its primary winding and 100 turns on its secondary winding. Find:

(1) The turns ratio;
(2) The secondary voltage if the input voltage to the primary winding was 240 V (ignore losses).

Now

$$\frac{\text{primary turns}}{\text{secondary turns}} = \text{turns ratio}$$

So that

$$\frac{1500}{100} = 15$$

Therefore

$$\text{turns ratio} = 15:1$$

This means that the output voltage will be 15 times smaller than the input voltage, so that:

$$\frac{240}{15} = 16 \text{ V}$$

In practice, transformers are often manufactured with more than one set of

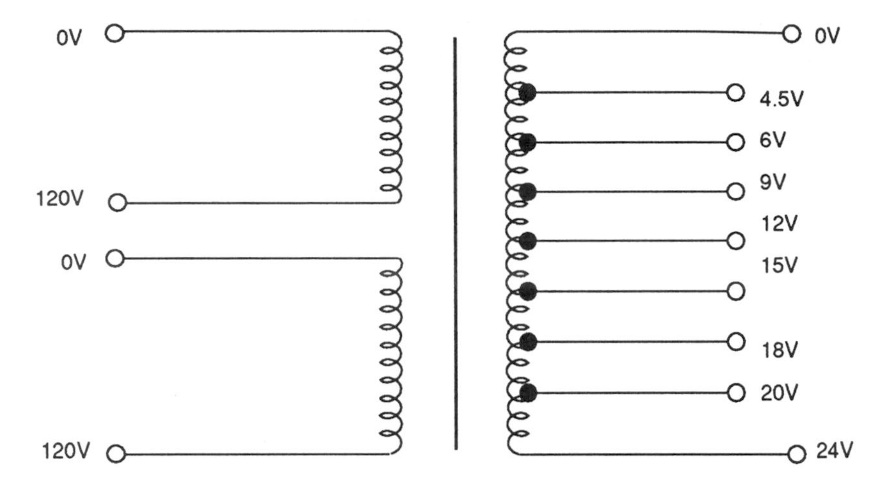

Fig. 2.3 Transformer tappings.

windings on the primary and secondary sides. Modern thinking is to produce equipment that is as versatile as possible. One way of doing this is to have the primary side wound with two sets of windings. These can be connected in series for 240 V working, or parallel for 120 V working. The secondary windings are often *tapped* to provide several different voltages on the output side of the transformer. The *taps* are simply connections on the secondary windings at intervals calculated to provide the range of voltages required (see Fig. 2.3).

2.18 Types of transformer

Air-core transformers These have their coils wound on a non-magnetic former. This type of transformer is not used as a power transforming device, since there are many magnetic losses due to leakage between the windings. These transformers are, however, widely used in communications equipment where they provide high selectivity over particular bandwidths.

Iron-core transformers These have separate primary and secondary coils (double wound) that are wound onto a specially shaped core made up of lots of iron plates or stampings (more often silicon steel in recent times). Although these are sandwiched together, each one is insulated from the other. The plates are called laminations and help to reduce eddy currents flowing in the core, which saves energy and reduces heat (see Information Sheet No. 2K.) Their main use is for *stepping up* or *stepping down* voltages, although they are useful as isolating transformers as there is no direct electrical connection between primary and secondary coils – only a magnetic one.

Ferrite-core transformers These find a variety of uses in electronics and are particularly suitable for output transformer service in switch-mode power

Information sheet No.2.K.

A typical transformer

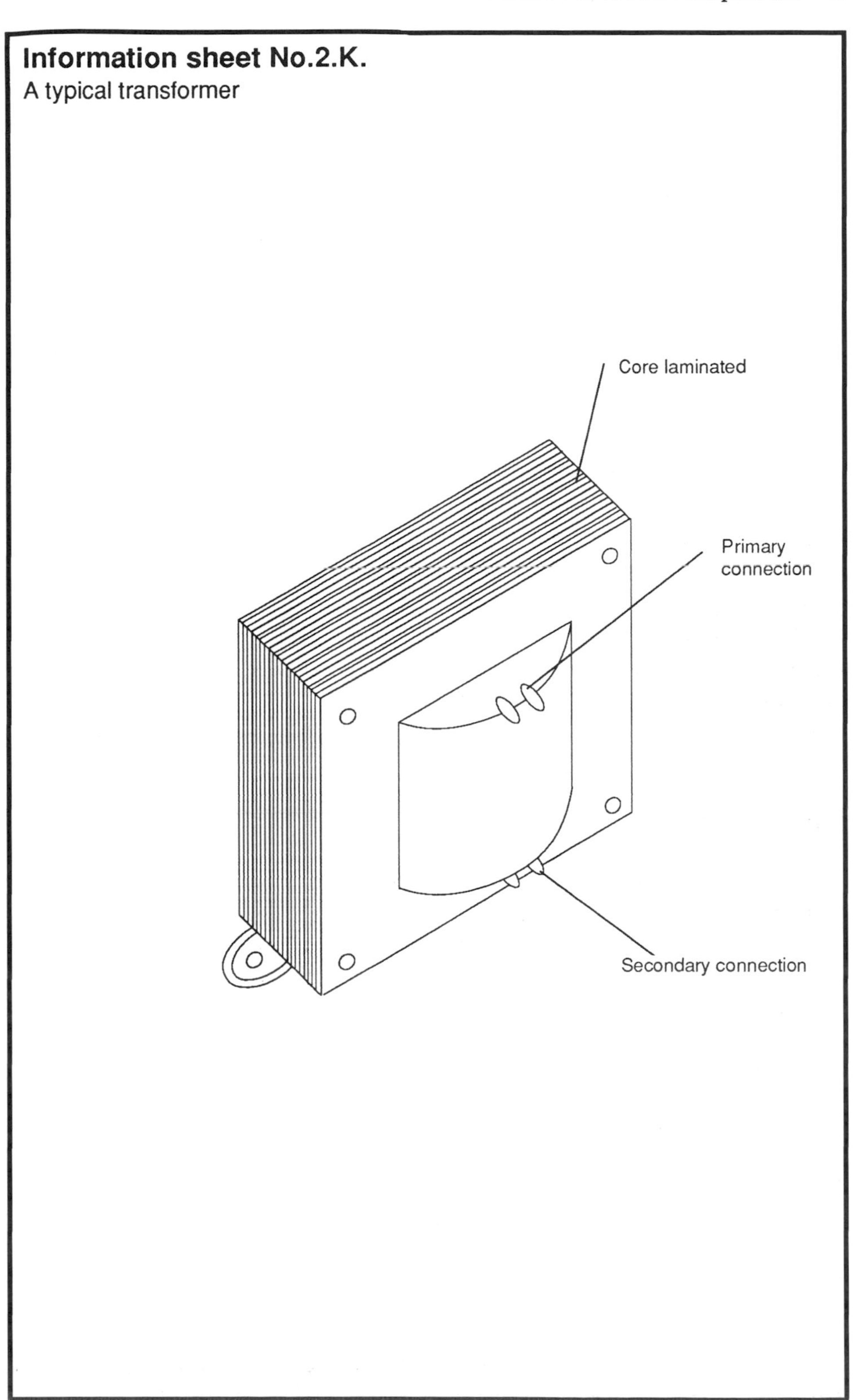

supplies and other high-frequency applications where high operating flux and low core losses have to be maintained even at elevated temperatures.

Toroidal transformers These are mains operated transformers of circular construction and are superior to the double-wound types. They have low magnetic interference, are small in size and light in weight. Care should be taken when mounting to ensure that both ends of the fixing bolt do not come into simultaneous contact with either a metal chassis or framework thereby creating an effective 'shorted turn' and causing damage to the transformer windings. This is usually achieved by use of a Neoprene disc supplied by the manufacturer.

2.19 Transformer ratings

The power rating of a transformer is usually quoted as a VA rating. This is the product of the secondary a.c. voltage and the secondary a.c. current. For example, a transformer with a 12 V secondary output has a maximum current rating of 0.25 A. Its VA rating is:

$$12 \times 0.25 = 3 \text{ VA}$$

If the VA rating of a transformer is known, the maximum current can be found by dividing the VA rating by the output voltage. For example, a transformer with a secondary output voltage of 15 V and rated at 30 VA, would have a maximum output current of:

$$\frac{30}{15} = 2 \text{ A}$$

Equally, if the VA rating is also true of the primary winding, it is possible to find the supply current. This is done by dividing the VA by the primary voltage. If the primary voltage was 240 V with a 30 VA rating, for example, then:

$$\frac{30}{240} = 0.125 \text{ A or } 12.5 \text{ mA}$$

SUMMARY

Sections 2.17–2.19 have described in simple terms inductors and transformers used in electronic circuits. The following important points should have been learned:

- That electricians should be able to recognize and identify the different types of inductors and transformers;
- How inductors and transformers are used in electronic circuits;

- How to recognize BS 3939 symbols used for the different types of inductors and transformers on circuit diagrams;
- How inductors and transformers are rated;
- An understanding of the terms *primary* and *secondary* in relation to transformers and how to carry out simple calculations involving these;
- The advantages and disadvantages of the different types of inductor and transformer.

Test 2

Choose which of the four answers is the correct one.

(1) A resistor has four colour bands; the fourth band represents:

(a) the working voltage,
(b) the tolerance,
(c) the multiplier,
(d) the wattage.

(2) A capacitor is a device which will:

(a) store a charge,
(b) store a current,
(c) create a magnetic field,
(d) store magnetic flux.

(3) An inductor would usually have a core of:

(a) resin or boxite,
(b) copper or brass,
(c) silicon or germanium,
(d) air or ferrite.

(4) A transformer will step up or step down:

(a) a.c. voltages,
(b) a.c. and d.c. voltages,
(c) d.c. voltages,
(d) a.c. and d.c. currents.

(5) The colour band representing 100 000 on a resistor is:

(a) orange,
(b) yellow,
(c) green,
(d) blue.

(6) The unit of measurement for an inductor is:

(a) ohm,
(b) kelvin,
(c) farad,
(d) henry.

(7) Electrolytic capacitors:

(a) have close tolerances,
(b) are rectifying devices,
(c) are polarized,
(d) have high inductance.

(8) A transformer's power rating is usually given in:

(a) kilovolts,
(b) volt-amps,
(c) newton-metres,
(d) microamps.

(9) The fourth colour band on a capacitor indicates the:

(a) working temperature,
(b) year of manufacture,
(c) working voltage,
(d) tolerance percentage.

(10) A resistor is marked 5K8F. What is its value and tolerance?

(a) 580 kΩ 1%,
(b) 5.8 kΩ 1%,
(c) 5.8 kΩ 2%,
(d) 580 kΩ 2%.

Chapter 3
Active Components

By definition, a component is said to be active if it delivers signal power to its surroundings. By *surroundings* we mean the circuit in which the component resides.

3.1 The junction diode

Diodes are very useful components that appear in great numbers in electronic circuits. They are semiconductor devices that have the curious property of only allowing current to flow through them in one direction, and are the electronic equivalent of a one-way valve. The symbol used for diodes is shown on Information Sheet No. 3A. The junction has two terminals called the anode and the cathode and the arrowhead on the symbol indicates which way the current will flow through them. As we will see, the current flows from the anode terminal to the cathode terminal and not the other way round.

If we suppose that sections of both the p-type and n-type semiconductor materials discussed in Section 1.4 are fused together, then an area exists where some of the electrons from the n-type material have moved into the p-type material. Similarly, some holes of the p-type material appear in the n-type material. This exchange of electrons and holes from one region to the other happens at the instant of manufacture, and creates an area where electrons and holes have combined. This is called the depletion layer and can be seen in Fig. 3.1.

Let us now connect a battery across the ends of the junction in the configuration shown in Fig. 3.2. Since like charges repel, the holes of the p-type material appear to be repelled by the positive side of the battery and move towards the negative side of the battery. The electrons of the n-type material are repelled by the negative side of the battery and move towards the positive side of the battery. The net effect of this is to reduce the depletion layer, and current flows in a forward direction. For silicon devices the voltage required to remove the depletion layer is in the order of 0.7 V and for germanium devices 0.3 V.

If we now connect the battery the other way round, as shown in Fig. 3.3, then, since like charges attract, the effect is that electrons in the n-type material are attracted to the positive side of the battery and holes in the p-type material appear to be attracted towards the negative side of the battery. The net effect is to further enlarge the depletion area and no current flows.

The above characteristics of the junction diode can be illustrated by the simple experiment shown in Information Sheet No. 3B. If we assume conventional

Information sheet No.3.A.
Diode symbols

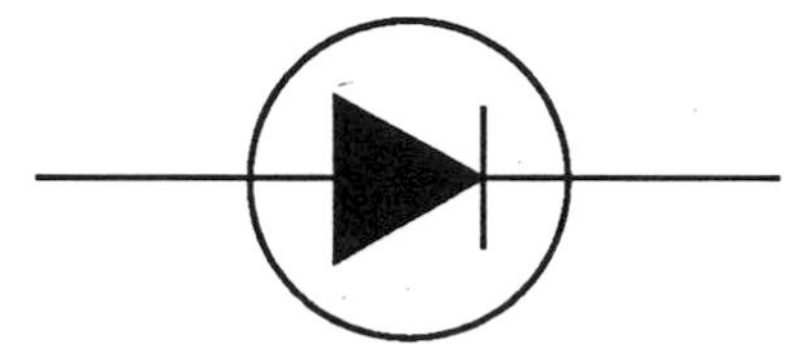

Symbol for junction diode

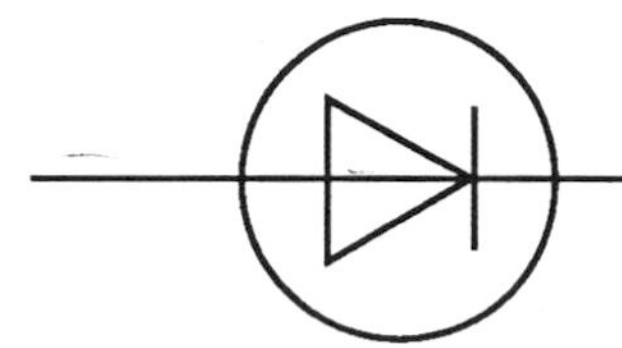

Alternative symbol

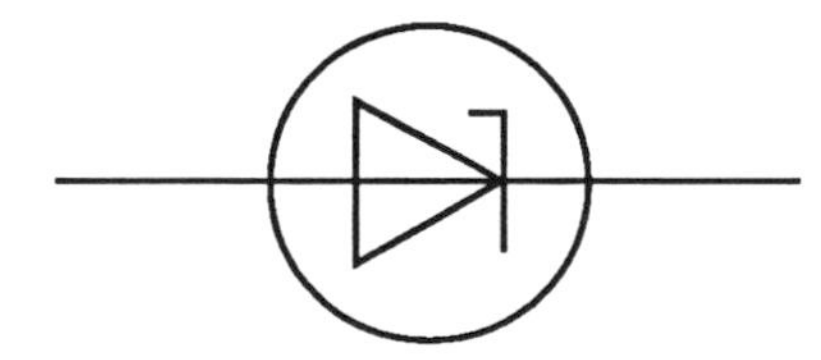

Zener diode

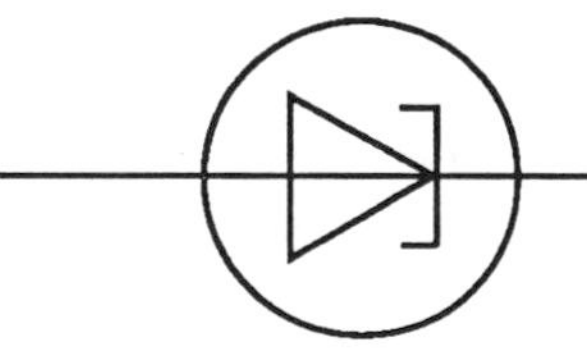

Tunnel diode

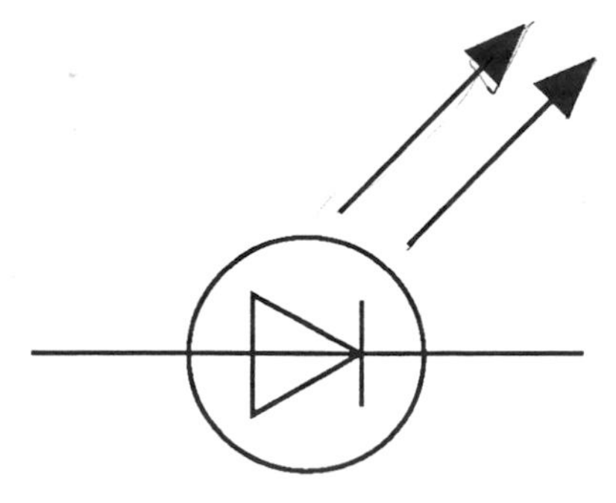

Light emitting diode

Note often seen without circle

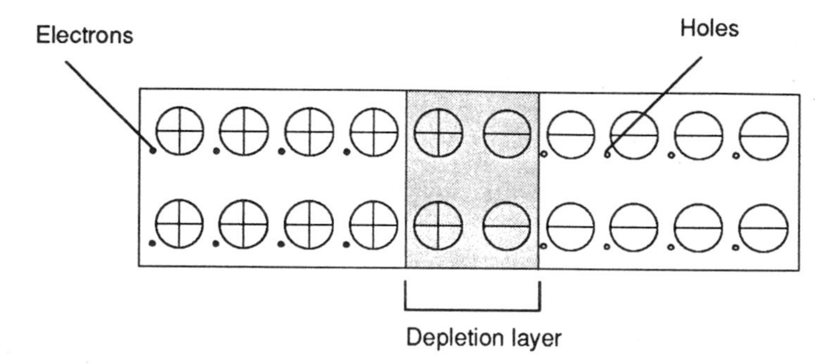

Fig. 3.1 Depletion layer.

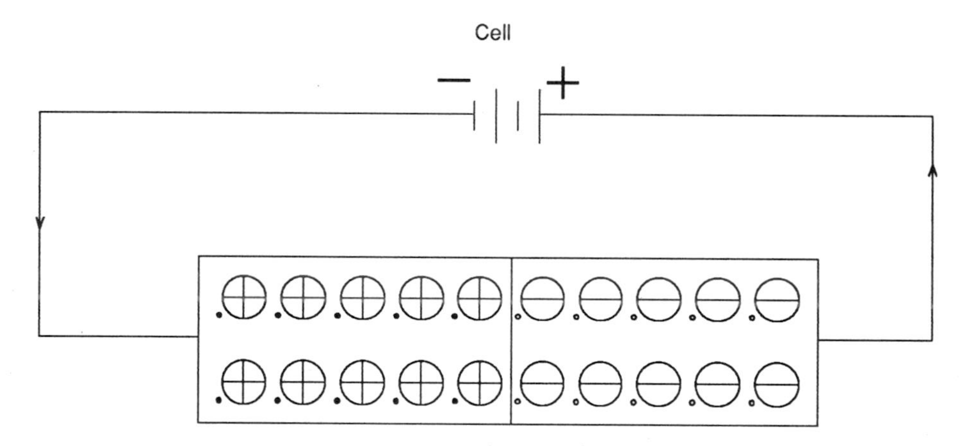

Fig. 3.2 Reduction of depletion layer.

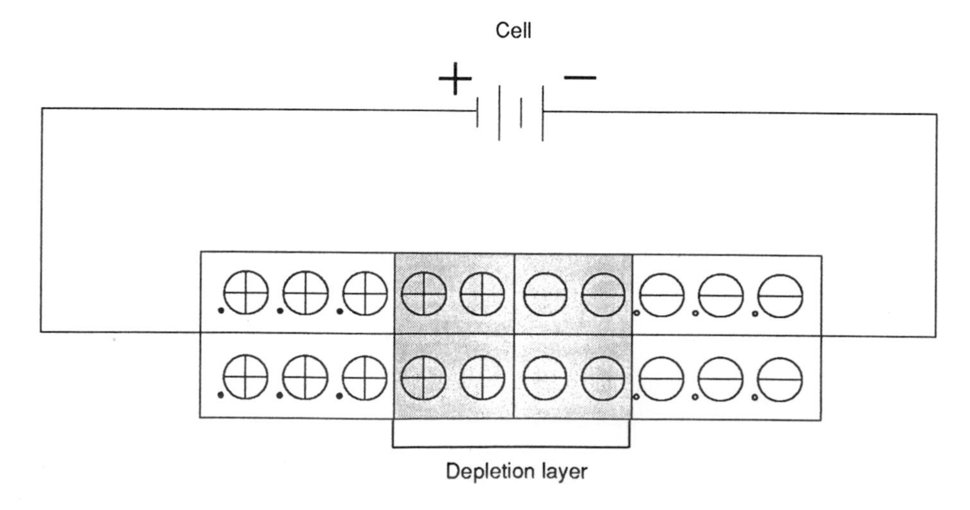

Fig. 3.3 Increased depletion layer.

Information sheet No.3.B.

Effects of diodes in lighting circuits

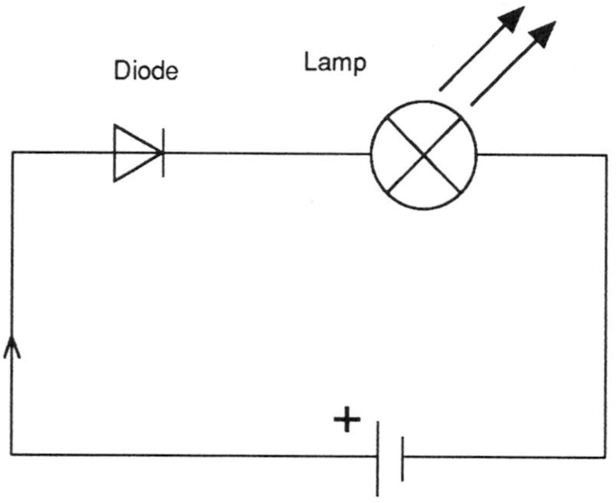

Diode

Lamp

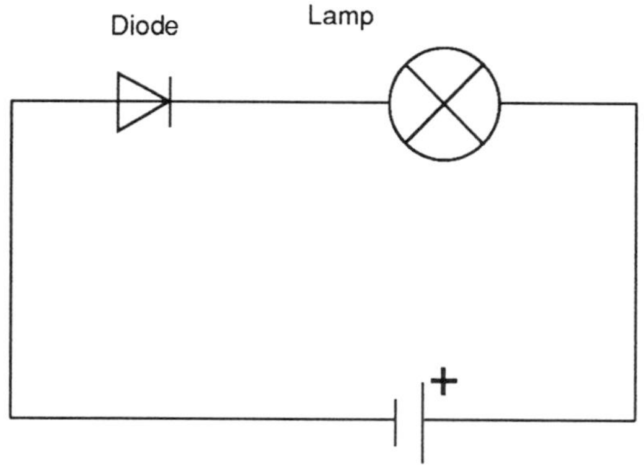

Diode

Lamp

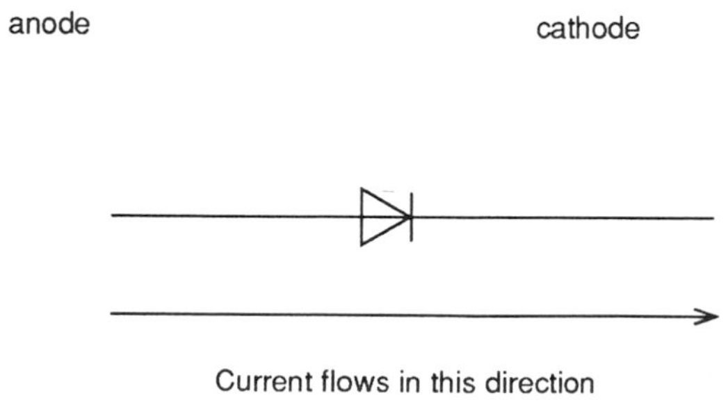

anode

cathode

Current flows in this direction

Information sheet No.3.C.

Current voltage characteristic of diodes

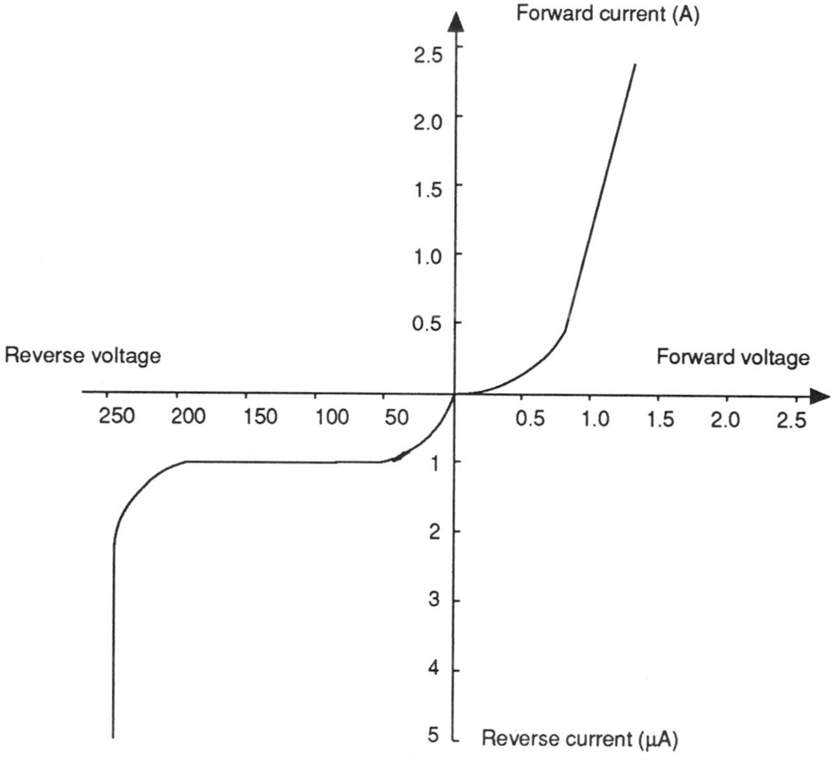

Forward current (A)

2.5
2.0
1.5
1.0
0.5

Reverse voltage

Forward voltage

250 200 150 100 50

0.5 1.0 1.5 2.0 2.5

1
2
3
4
5

Reverse current (μA)

current flow, the lamp in drawing (a) will be illuminated as the depletion layer is reduced and current flows. When connected in this way the device is said to be forward biased, i.e. the current flows in a forward direction.

If the device is connected as in drawing (b) then the depletion layer is increased and blocks the flow of current, and the lamp will not be illuminated. When connected in this way the device is said to be reverse biased, i.e. no current flows in the circuit.

In practice, there will be a maximum current that a diode can pass when forward biased; this is referred to as the forward current (I_f). When the diode is reverse biased there will be a very small leakage current; this is referred to as the reverse current (I_r). Reverse leakage current is temperature dependent, the leakage current approximately doubling for every 10°C rise in temperature. For all intents and purposes reverse leakage current can be ignored.

With diodes of forward current ratings of greater than 6 A, heat produced by the device becomes a problem. For this reason, high power diodes are made with a housing that can be bolted to a heat sink to dissipate the heat.

Just as there are forward and reverse current ratings, so there are forward voltages (V_f) and reverse voltages (V_r). If the voltage in the reverse direction is increased beyond a certain critical point, the diode will break down. The breakdown voltage is always given by the manufacturers for these devices and is often referred to as the peak inverse voltage (PIV) or even reverse repetitive maximum voltage (Vrrm).

Peak inverse voltages may be as low as 30 V for germanium to as high as 10 000 V for certain silicon diodes. It is important that these are not exceeded as this will result in the complete destruction of the device, rendering it useless. Information Sheet No. 3C shows the current voltage characteristics for a typical semiconductor diode.

3.2 Types of diodes

The numerous different types of diode, all having different characteristics from one another, have one thing in common: they will only conduct if the polarity of the supply is correct. That is why they are sometimes referred to as *polarity conscious devices*.

Signal diodes These are ideal for use in a wide range of low power applications including high speed switching, detectors, demodulators, mixers, etc. They vary from devices rated at 4 V Vrrm, and an I_f of 30 mA, used in ultra-high-frequency (UHF) mixers, to 350 V Vrrm with an I_f of 400 mA, for a general purpose fast switching diode. Identification of the axial lead types is by way of a coloured band round the cathode end of the device, other types of package resort to printing details (sometimes difficult to decipher) on the side of the component.

Power diodes Power diodes of the smaller axial lead types can range from devices rated at 50 Vrrm and an I_f of 1 A, to ones rated at 400 Vrrm and an I_f of 3 A. They are much used for higher current switch mode power supplies and similar fast rectifier applications. Identification once again is by a coloured band on the cathode end of the component.

Stud-mounting diodes As we learned earlier, these devices are used for power diodes carrying large currents. They are available with either anode or cathode stud-mounting so test with a meter to verify which is which. The connection at the other end can be either a solder terminal or have a flexible lead supplied complete with connection lug. Heat dissipation is the main problem for these diodes but with the right heat sink they can be rated as high as 1000 Vrrm with an I_f of 240 A. Applications include high power switching, rectification of a.c. power supplies for battery charging, etc. The subject of rectification is dealt with in Chapter 6.

Zener diodes These look very much like ordinary diodes from the outside; in fact, some of the outlines are identical to junction diodes. The BZY88 series of 400 mW zeners, for example, have the DO7 diode outline and only the identification of the type number will help. If the number is unreadable it would be wise not to make assumptions; a lot of damage could be caused if you were wrong. The cathodes of many of the zener diodes are marked by a coloured band in the same way as ordinary diodes. The circuit diagram symbol is also similar (see Information Sheet No. 3A). In cases where the identification of a component is doubtful, reference should be made to the circuit diagram for the equipment concerned.

The zener diode is widely used as a voltage reference device in power supplies and is often referred to as a *reference diode* or *reference voltage* component. To understand this, we must first compare the characteristics of the ordinary diode with those of the zener diode.

With the ordinary diode, if the voltage is reversed and allowed to increase beyond a certain critical point (Vrrm), then the device will break down and current will flow in both directions. This current can damage an ordinary diode but a zener diode is designed to operate in the reverse breakdown region.

We can make use of the above phenomenon by deliberately connecting the zener diode the opposite way round in circuit, i.e. the cathode positive and the anode negative. Connected like this, the reverse current will be zero until the reverse voltage reaches a certain predetermined level (V_z). At this point the device will 'turn on' and allow current to flow. Zener diodes are designed to break down in the reverse direction, once a certain voltage (called the breakdown voltage) has been reached, without damage to the diode.

Zener diodes have a higher resistance to current flow, in the reverse direction until the breakdown voltage is reached. At this point the diode breaks down and current flows through the diode in the reverse direction. The zener diode does not obey Ohm's Law and the voltage across the diode will be the same as the breakdown voltage, no matter what the current through the diode. If the voltage across the device falls to below the breakdown voltage, the diode returns to its original high resistance state. In the forward direction it acts in the same way as a normal diode.

Good use is made of this feature of the zener diode in applications such as stabilizing power supplies and this will be discussed in Chapter 6.

A look at an electronic component catalogue will show that as well as the breakdown voltage, the power rating of the zener diode is also given. Zener diodes in the BZX79 series, for example, are available with breakdown voltages from 2.4 to 75 V with a power rating of 500 mW (0.5 W). The power rating must be taken into consideration, because from this the maximum forward current can be calculated.

EXAMPLE 3.1

A zener diode of 9.1 V has a power rating of 500 mW. What is the maximum I_f the device will handle?

Now $$P = I \times V$$

Transposing for I gives $$I = \frac{P}{V}$$

Therefore we can say $$I_f = \frac{W}{V_z}$$

So that $$I_f = \frac{0.5}{9.1}$$

$$I_f = 0.55 \text{ A or } 55 \text{ mA}$$

Light emitting diodes (LEDs) These are semiconductor diodes which emit light when current passes through them. This occurs in a very complex manner beyond the scope of this book; fortunately a detailed knowledge of LEDs is not needed in order to use them. They are available in four different colours: **red, yellow, orange** and **green**. Their physical construction is slightly different for each colour.

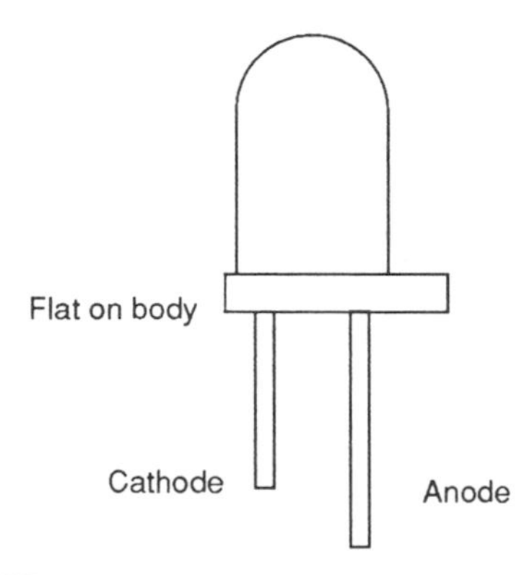

Fig. 3.4 Appearance of LED.

In operation, LEDs behave just like ordinary diodes in that they allow current to flow in one direction, but the similarity ends here because their appearance is quite different. The typical construction of a LED is shown in Fig. 3.4.

SUMMARY

Sections 3.1 and 3.2 have described in simple terms types of diodes used in electronic circuits. The following important points should have been learned:

- That electricians must be able to recognize and identify the different types of diodes;
- How to use diodes in electronic circuits. A simple concept of their principle of operation should also have been gained;
- How to recognize BS 3939 symbols used for the different types of diodes on circuit diagrams;
- The meaning of the expressions *forward bias*, *reverse bias* and *breakdown voltage*;
- The meaning of the term 'reference diode' in connection with the zener diode and the importance of establishing the power rating of the device;
- The advantages and disadvantages of the different types of diode.

3.3 The transistor

The transistor is responsible for the great advances in electronics that we see today. Developed in 1948 by Nobel prize winners, Bardeen, Brittain and Shockley, transistors have made an impact that few could have foreseen. They are semiconductor devices and a look at a manufacturer's catalogue would show that there are numerous different types available. Modern transistors are almost certain to be of the silicon type; germanium transistors like the germanium diode are little used.

3.4 Transistor types

Transistor devices can be classified into two main groups:

(1) *Bipolar transistors* – current operated devices;
(2) *Field effect transistors* (FETs) – voltage operated devices.

Information sheet No.3.D.

Circuit symbols for bipolar transistors

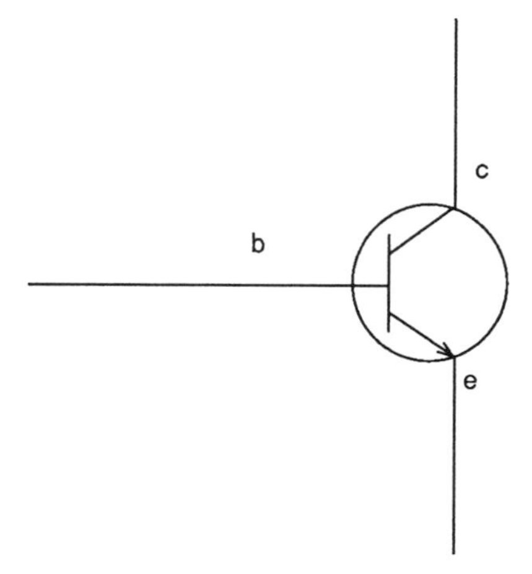

n.p.n. transistor

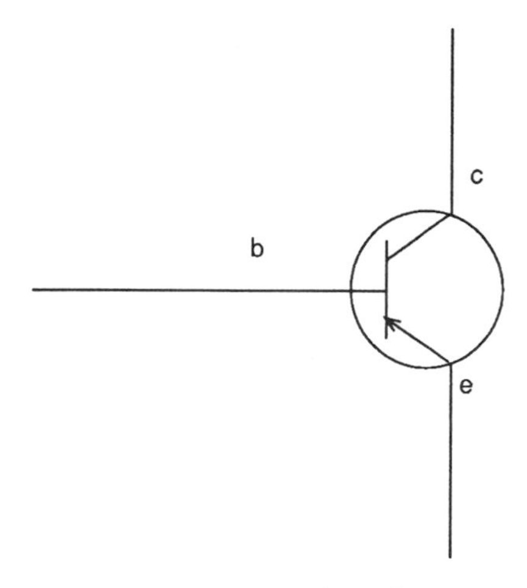

p.n.p. transistor

When used in circuit transistors are generally used in one of two different ways:

- Operating as an electronic switch
- As an amplifier to increase electrical signals.

Bipolar transistors

In its simplest form, this transistor could be considered to be two junction diodes joined back-to-back. The sandwich formed in this way could have the n-type material in the middle with p-type material on either side (pnp), or alternatively it could have the p-type material in the centre and n-type material either side (npn). The two are very similar, with the npn-type mainly conducting electrons and the pnp-type conducting mostly holes. Lead-out wires are attached to each of the three regions, the central region being called the base (b), one of the outside regions being called the collector (c) and the other being called the emitter (e). The circuit symbols are shown on Information Sheet No. 3D. The direction of the arrow on the emitter shows the direction of flow of conventional current for each type.

The npn-type silicon transistors are the easiest to construct, so naturally they are the most popular in use, both for amplification and switching. We shall therefore concentrate our studies on the npn-type transistor.

Using the bipolar transistor as an amplifier

We know from our previous work on the junction diode that a pn junction can be either forward biased, allowing current to flow, or reverse biased, preventing current flow (ignoring leakage current). Remember, the bipolar device is two np junctions in one and is current operated.

For the transistor to operate as an amplifier it is necessary for two things to happen at the same time:

(1) That the emitter and base junction is forward biased;
(2) That the collector base junction (perhaps surprisingly) is reverse biased.

Information Sheet No. 3E shows this together with the overall transistor biasing.

Under these conditions as the emitter-base junction is forward biased, conduction is mainly by means of free electrons passing from the emitter material into the base; this is because there are few positive holes in the p-type material of the base which can pass into the emitter because the base is only lightly doped.

When electrons from the emitter enter the base the chances of their being neutralized by a hole are small. This is because, firstly, the base is very thin (about 0.025 mm) and secondly because the concentration of holes is quite low. Most of the electrons entering the base from the emitter are therefore drawn through the depletion layer into the collector by the electric field which exists across the

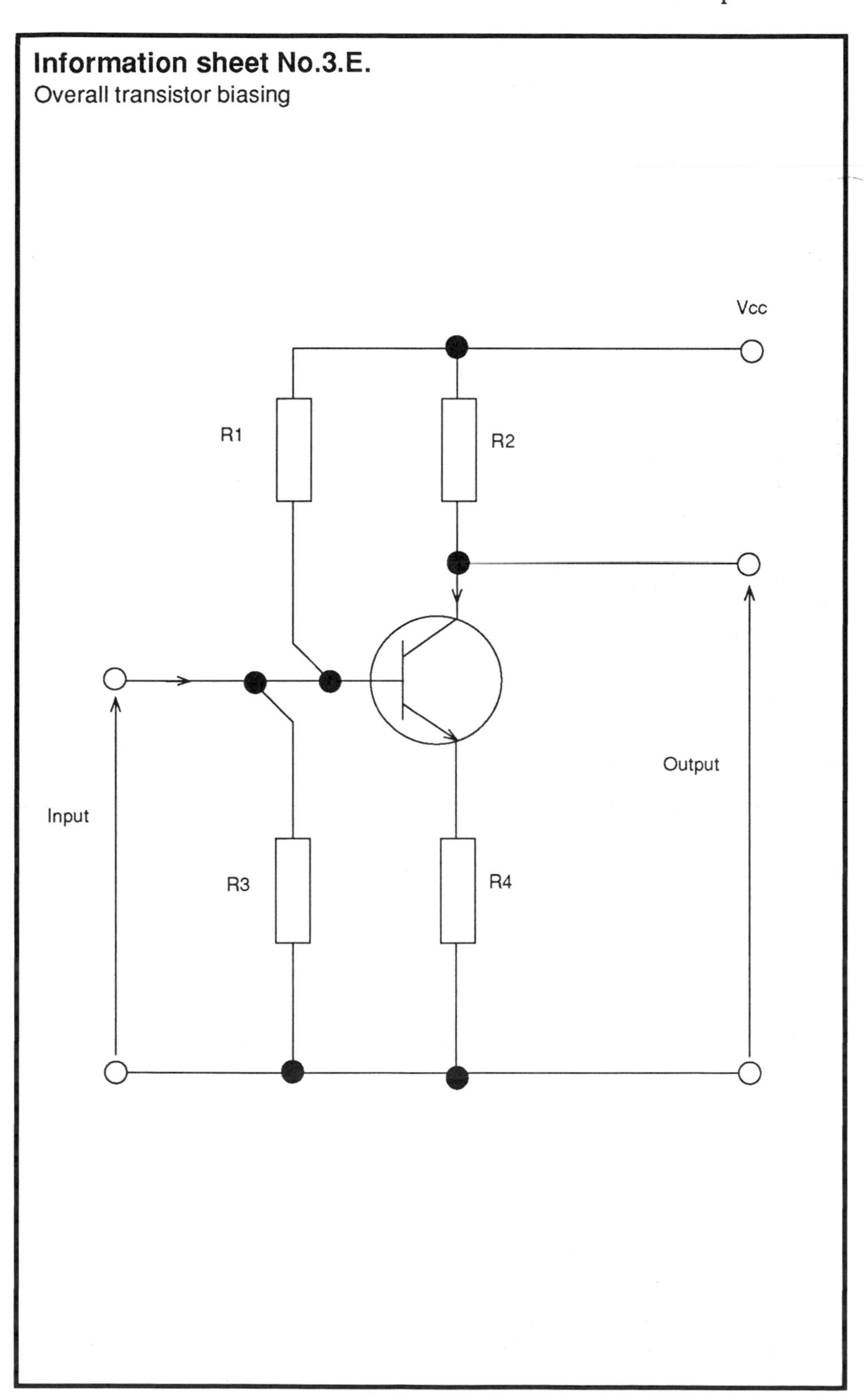
Information sheet No.3.E.
Overall transistor biasing
Vcc
R1
R2
Input
Output
R3
R4

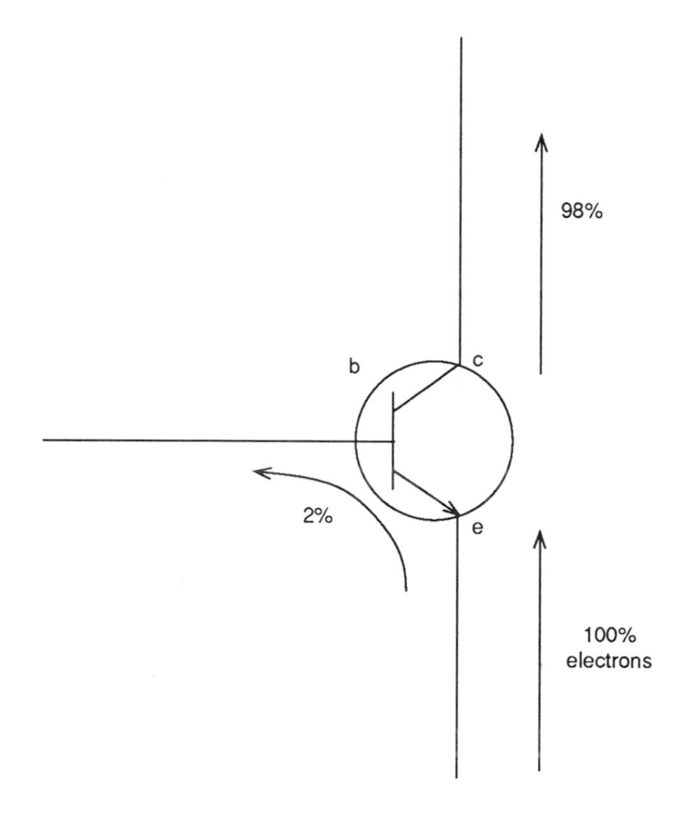

Fig. 3.5 Electron flow through an npn transistor.

collector–base junction. As electrons move to the collector, they are quickly replaced by others which enter the base from the emitter. In most transistors about 98% of the electrons from the emitter reach the collector and only 2% combine with the holes in the base (see Fig. 3.5).

The number of electrons in the base region is determined by the value of the current from the emitter to the base region and this controls the number of electrons reaching the collector and hence the collector current. We can see that as the base current increases so the collector current increases in proportion to it, so this simple transistor circuit provides current amplification and the collector current is many times greater than the base current.

EXAMPLE 3.2

Input base current, 20 μA; Output collector current, 1 mA.

This gives a current gain of:

$$\frac{1\ \text{mA}}{20\ \mu\text{A}} = 50.$$

Using the bipolar transistor as a switch

When transistors operate as a switch, an analogy with an electromechanical switch (a relay) can be drawn. A transistor that is fully *on* behaves like an electromechanical switch that is closed.

For the electromechanical switch to operate, a current must be supplied to its coil for the switch contacts to close. For the transistor to operate we need to provide enough current for the transistor to saturate which means that an increase

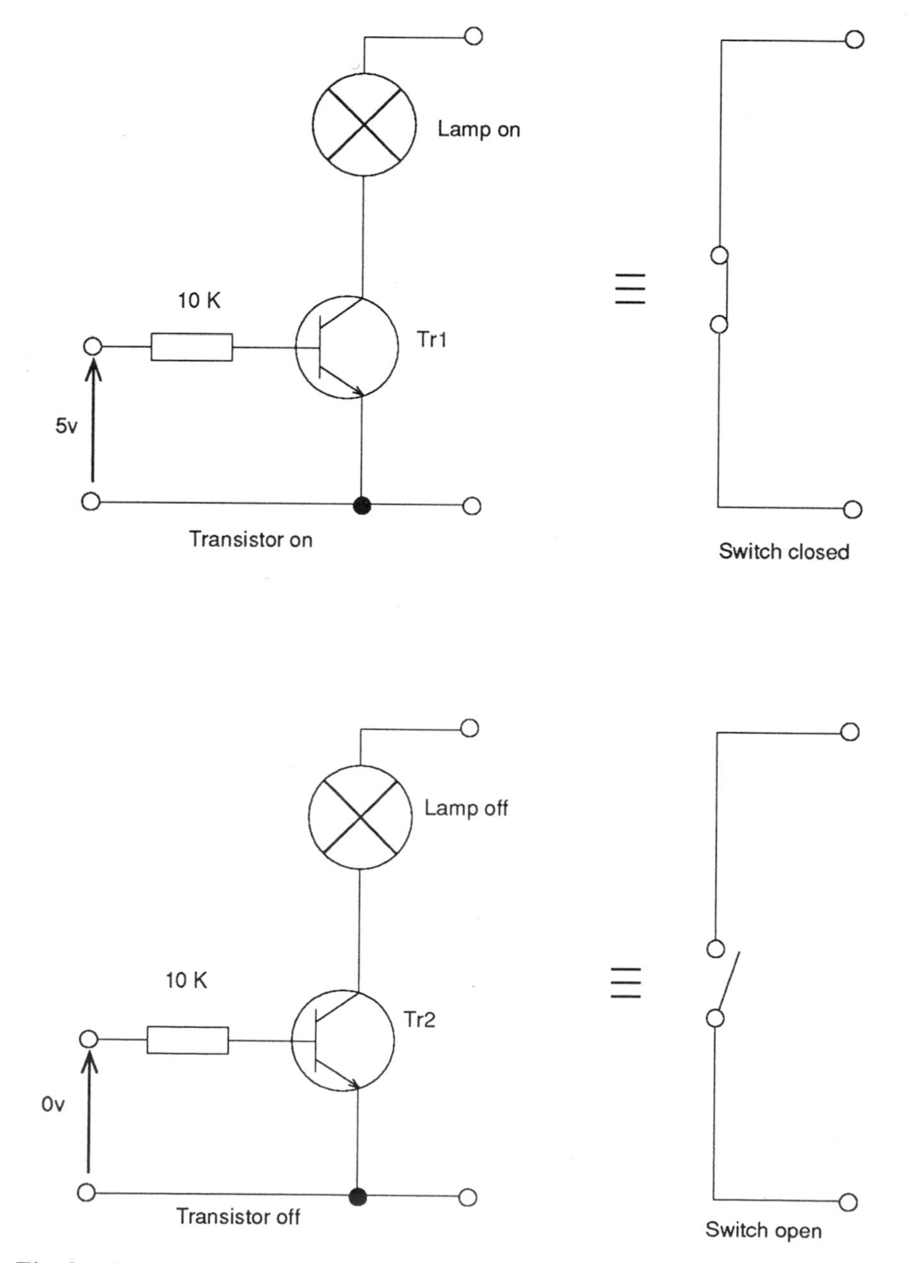

Fig. 3.6 The transistor as a switch.

in base current will not significantly increase the collector current.

For the electromechanical switch to be in the *off* position we no longer supply a current to the coil and the contacts open. When the transistor is *off*, it is just like a switch that is open and for this condition no base current is supplied to the transistor.

Figure 3.6 shows a transistor in circuit with a lamp. By closing the switch, the base-emitter junction will be forward biased and current flowing through this junction causes current to flow across the collector-emitter junction as indicated by the arrow, and the lamp is lit.

Negative electrons moving from the emitter through the base into the collector, as described above, are equivalent to a conventional current flow in the opposite direction from collector to emitter. The arrow on the circuit diagram indicates conventional current flow, which is in the opposite direction to electron flow. Although we have described the operation of the transistor with the idea of electron flow in mind, it is often necessary to consider the working of the transistor circuit in terms of conventional current flow.

Field effect transistors

Field effect transistors (FETs) are important components in the field of modern electronics and no book about electronics would be complete without their mention.

FETs are semiconductor devices and the different types make use of both p-type and n-type materials in different ways, in accordance with the characteristics required of the component. Like the bipolar transistor, the FET is a three-electrode device. These electrodes are known as the source, gate and drain and correspond to the emitter, base and collector of the bipolar transistor.

The bipolar transistor, as mentioned earlier, is a current-operated device, the current between the collector and emitter being controlled by the amount of base current. We have seen that the conduction processes taking place make use of both types of charge carrier: namely the electron and the hole. The FET, however, is a voltage-controlled device, the amount of current allowed to flow between the source and the drain being controlled by the gate voltage. It is, in effect, a unipolar device making use of only one type of charge carrier to carry most of the current. This gives it a big advantage over the bipolar device as the high gate resistance that FETs possess, which is in the order of gigaohms (GΩ) or even teraohms (TΩ), means that they do not load the source of the signal - a distinct possibility for bipolar devices.

There are two important types of FET:

- The junction FET (JFET), sometimes known as the junction gate FET (JUGFET);
- The metal oxide semiconductor FET (MOSFET), sometimes known as the insulated gate FET (IGFET).

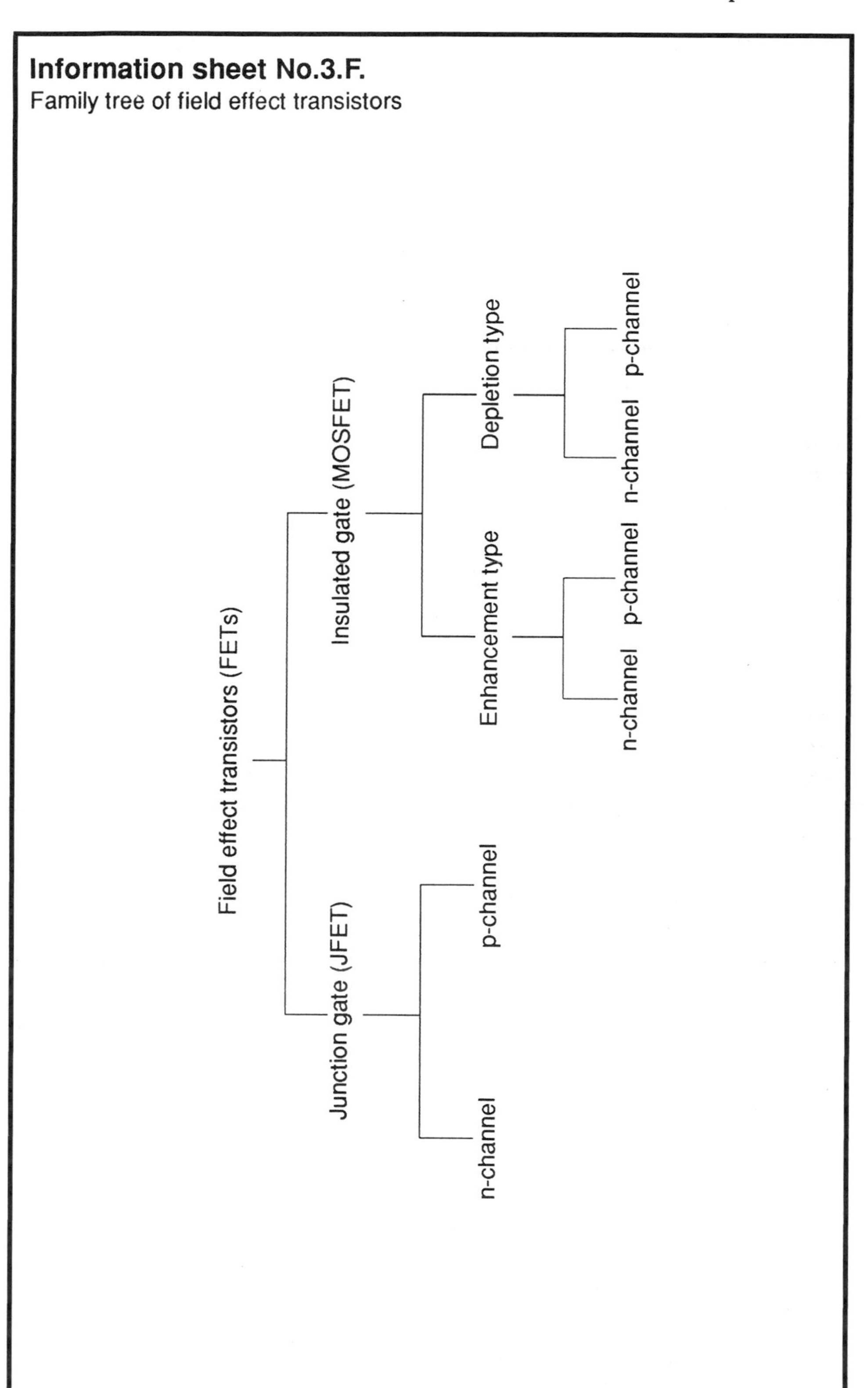
Information sheet No.3.F.
Family tree of field effect transistors
Field effect transistors (FETs)
Junction gate (JFET)
Insulated gate (MOSFET)
n-channel
p-channel
Enhancement type
Depletion type
n-channel
p-channel
n-channel
p-channel

Information sheet No.3.G.

Construction of a JFET

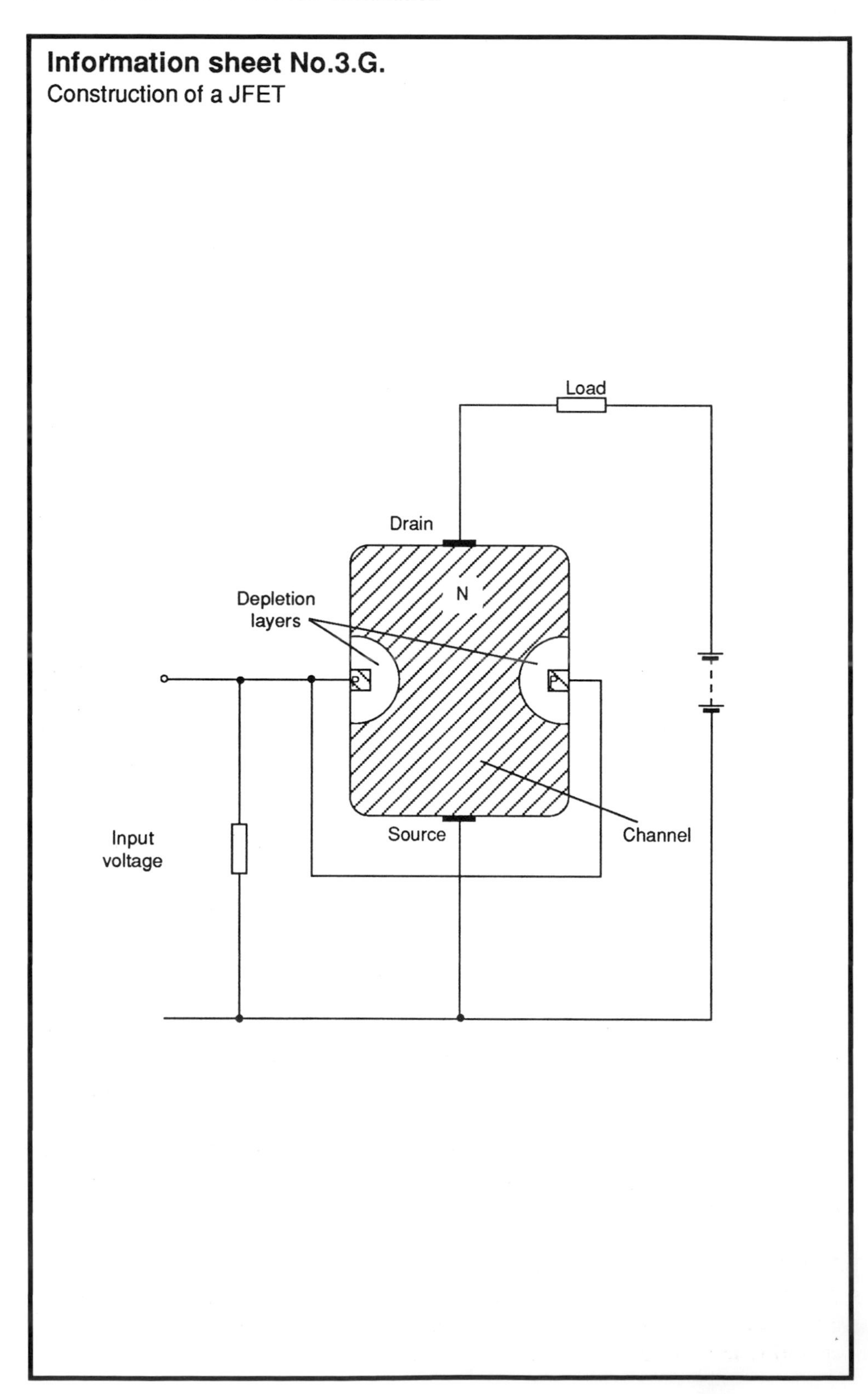

They can be categorized as follows:

(1) Junction FET (JFET)
  (a) n-channel, normally conducting, but reverse bias reduces conduction;
  (b) p-channel, normally conducting, but reverse bias reduces conduction.

(2) Metal oxide semiconductor FET (MOSFET)
  (a) n-channel enhancement type, normally not conducting, but forward bias causes conduction;
  (b) p-channel enhancement type, normally not conducting, but forward bias causes conduction;
  (c) n-channel depletion type, normally conducting, but reverse bias reduces conduction;
  (d) p-channel depletion type, normally conducting, but reverse bias reduces conduction.

Information Sheet No. 3F shows a family tree of FETs.

The Junction FET (JFET) A block of silicon n-type or p-type material called the substrate has connections made at either end; these are called the source and the drain. To form an n-channel the block is surrounded lengthwise by p-type material and to form a p-channel the block is surrounded by n-type material, as shown on Information Sheet No. 3G. A depletion layer is formed on either side of the metallurgical junction resulting from the outer material making contact with the material of the channel. An electrode is connected to the outer layer and this is known as the gate.

If a battery were to be connected across the source and the drain of an n-channel type, then electrons would flow from the drain down the channel and out through the source (or from the source to the drain, these two being interchangeable).

Now look what happens if the gate is made negative with respect to the source. The two p-n regions form reverse bias diodes and so the depletion area around them extends out into the channel and reduces the amount of channel that the electrons can flow through. If the gate is made sufficiently negative the channel will become virtually closed. This is known by the expression *pinch off*. The conduction between the source and drain is entirely due to majority carriers in the channel: for the n-channel type this is due to electrons; for the p-channel type it is due to holes.

If the input to the gate is a signal which is varying, then this will cause corresponding changes in current between drain and source, and so the basis of an amplifier is formed.

The metal oxide semiconductor FET (MOSFET) While both the JFET and the MOSFET devices are unipolar devices, i.e. they rely on one type of carrier for conduction, there are fundamental differences in their construction and operating principles.

Information sheet No.3.H. (a)

N-channel enhancement type MOSFET

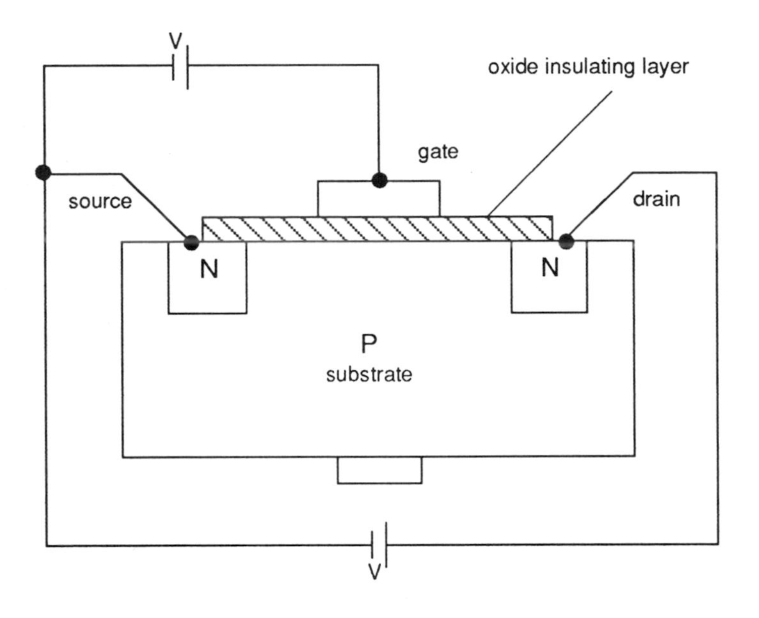

The JFET, as its name implies, relies very much on normal junction behaviour; the channel conduction is controlled by the depth to which the depletion layer extends out into the channel. The MOSFET, on the other hand, relies on the existence of what is known as an *inversion layer* of charge carriers.

There are two types of MOSFET:

(1) The *enhancement-type*, in which the conducting channel is made more conductive by gate biasing;
(2) The *diffused-channel-depletion-type*, in which the conducting channel is depleted of charge carriers by gate biasing.

Information Sheet No. 3H(a) shows a cross-sectional view of the *enhancement-type MOSFET device*. Here a piece of silicon p-type material, known as the substrate, has n-type regions at each end and these form the source and the drain electrodes. The gate consists of a metal electrode (aluminium) formed on the top surface of the substrate and isolated from it by an insulating layer of oxide (silicon dioxide).

With no potential at the gate, there is a very high resistance between source and drain as there are effectively two p-n junctions back-to-back. The gate and the substrate act like the plates of a capacitor, with the insulating layer of oxide in between them forming the dielectric. If the gate is now made positive with respect to the source, the capacitive effect of the device draws electrons to a layer under the gate known as the inversion layer. The inversion layer serves as a conductive path between source and drain and the ability to conduct will be enhanced as a result.

Information Sheet No. 3H(b) shows the *diffused-channel-depletion-type MOSFET*. The main difference between this and the enhancement-type is that a permanent n-channel is diffused between the source and the drain.

With no bias applied, current flows between source and drain. When the gate is made negative with respect to the source, the n-channel electrons are repelled and this creates a depletion layer. The channel between source and drain becomes narrower or depleted and so less current flows; the device is said to be working in the depletion mode.

The depletion-type MOSFET shows similar characteristics to the JFET under the above conditions. The two devices differ, however, when the gate potential is made positive with respect to the channel:

- With the JFET, the p-n junctions become more forward biased and majority current flows;
- With the depletion-type MOSFET more negative charge is induced to the n-channel, so enhancing the conduction.

This means that the depletion-type MOSFET may be used in both the depletion mode and the enhancement mode.

The input resistance of the MOSFET is very high. This is due to a large extent

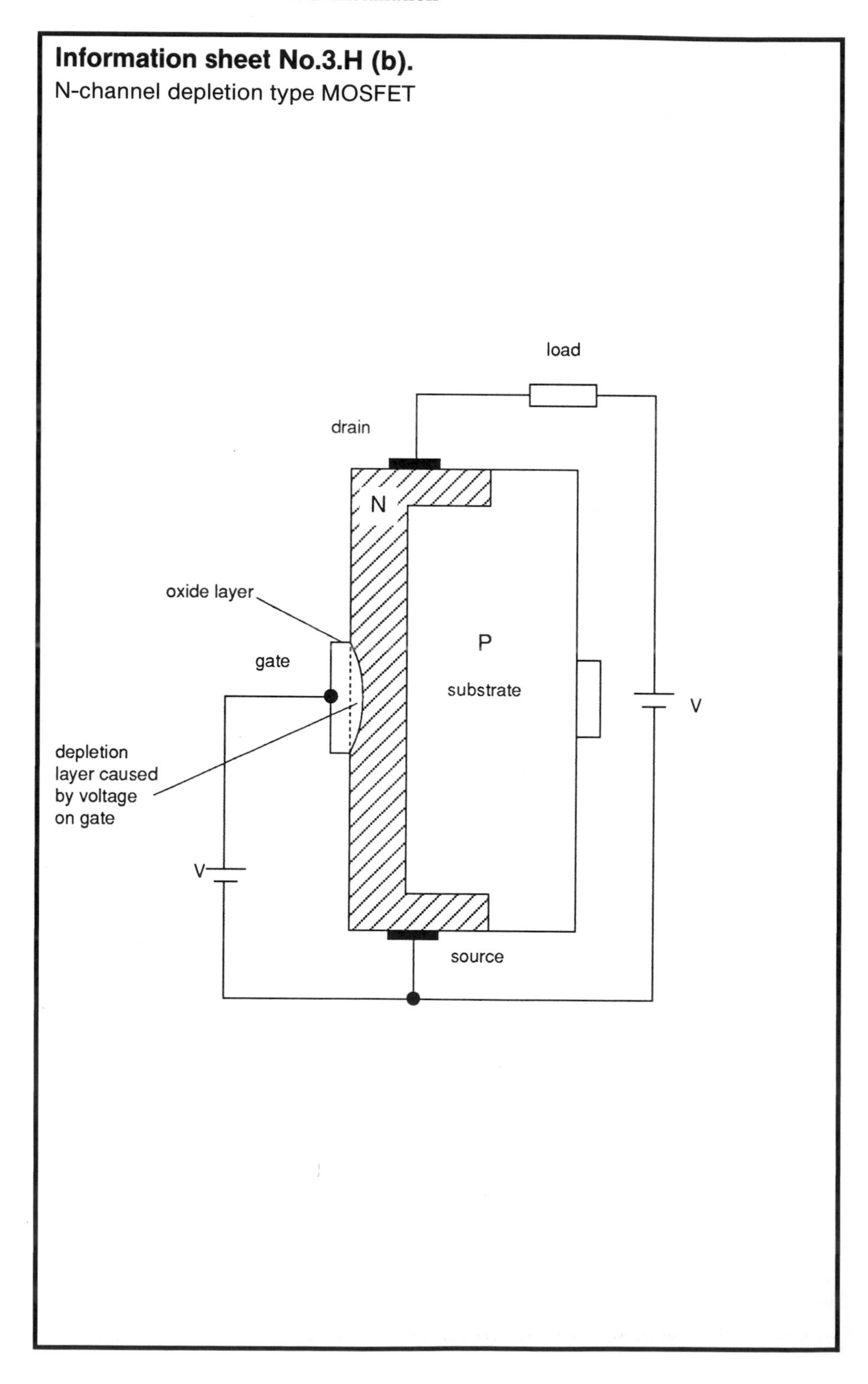
Information sheet No.3.H (b).
N-channel depletion type MOSFET
load
drain
N
oxide layer
gate
P
substrate
V
depletion
layer caused
by voltage
on gate
V
source

Information sheet No.3.I.

Symbols for thyristor

anode cathode

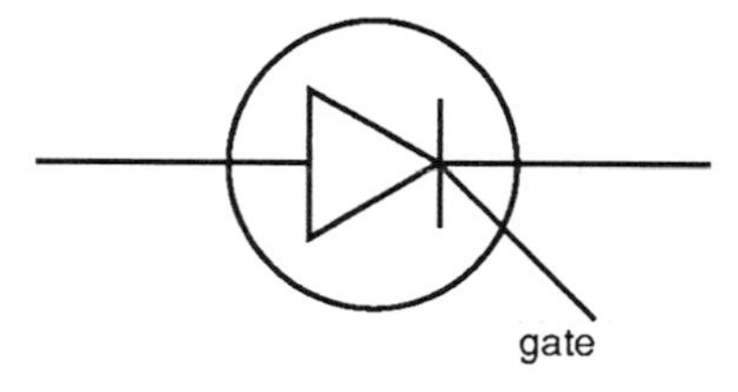

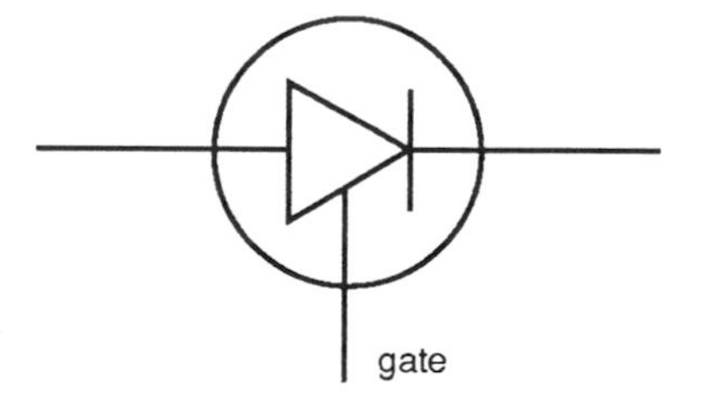

alternative symbol

to the insulating layer between the gate and source. The MOSFET is useful in switching applications.

3.5 Thyristors

The thyristor or silicon-controlled rectifier (SCR) is a high speed electronic switch which can handle large currents without arcing or excess heat and requires no maintenance. It is now replacing contactors and mechanical switches in many electrical applications. The circuit symbol is shown on Information Sheet No. 3I.

Thyristors are four-layer devices of the format pnpn, and have three terminals called the cathode, anode and gate. Under normal conditions the device will block currents in both the forward and reverse direction (unlike a conventional diode). To turn the thyristor on (make the thyristor conduct) in the forward direction, then, provided the anode is more positive than the cathode or gate, all that is required is the application of a signal to the gate terminal. However, removal of the gate signal will not turn it off again! To turn the device off, the current passing through it must be reduced until it is below the holding current for the device, or the anode voltage removed. The gate signal has no effect on the thyristor in the reverse direction.

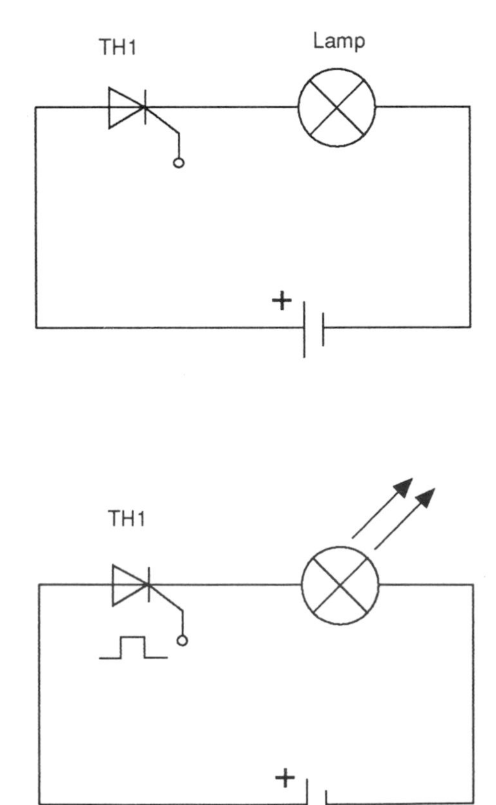

Fig. 3.7 Thyristor in a d.c. circuit.

Information sheet No.3.J.

Thyristor circuit for use in d.c. applications

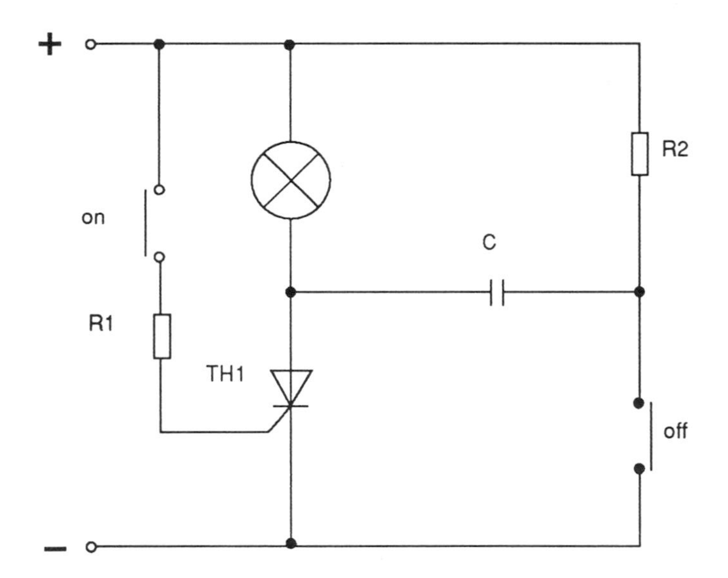

3.6 Thyristors in d.c. circuits

Consider the circuit in Fig. 3.7. The thyristor is off and so the lamp will not be lit. With the application of a gate signal the thyristor will be turned on and the lamp will light. Because the current from the battery is d.c., and always at a constant value (not alternating), the light will not turn off until either:

(1) The bulb filament breaks;
 or
(2) The battery goes flat.

Both these conditions will reduce the current through the device to below its holding current.

To use the thyristor for control purposes in a d.c. circuit, further circuitry must be included in order to reduce the current through the device to below that of the holding current. These circuits reduce the current for only a fraction of a second, but this is time enough to turn off the device. Information Sheet No. 3J shows a low voltage thyristor circuit for d.c. use.

3.7 Thyristors in a.c. circuits

The thyristor really comes into its own in a.c. circuits. By controlling the conduction time of the device over each cycle of the mains voltage, the power delivered to the load can be varied. There are two main methods of achieving this control:

(1) Phase control;
(2) Burst control.

Phase Control

It is possible to turn the device on at any point in the positive half cycle of the a.c. supply. The point at which the gate signal is applied (sometimes called the firing angle) controls the amount of current delivered to the load. Altering the firing angle will vary the point at which the thyristor conducts. Firing at the beginning of the wave will allow almost all the positive half cycle, whilst firing towards the end allows only a small fraction to be let through. This is shown on Information Sheet No.3K. This method of control is ideal for dimming lighting and varying the speed of electric motors.

A single thyristor will control the power delivered to a load from zero to half power. For full wave control, two thyristors are used, as shown on Information Sheet No. 3L.

Information sheet No.3.K.

Thyristor firing diagrams

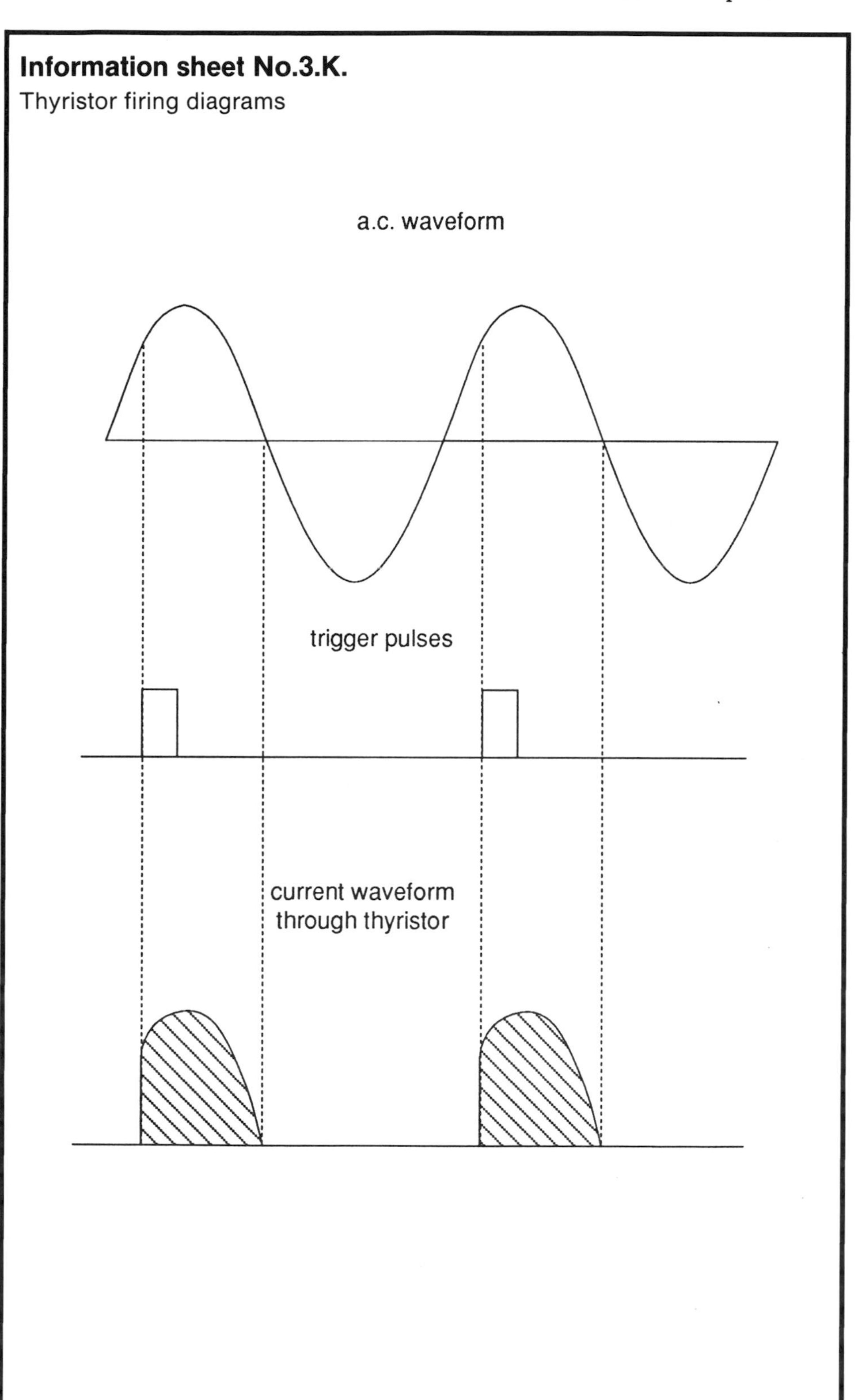

Information sheet No.3.L.

Full wave control using thyristors in an a.c. circuit

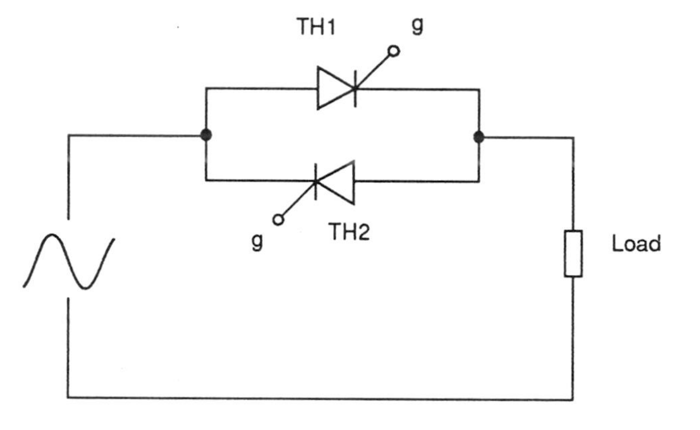

Burst Control

With burst control the gate signal is applied for a number of half cycles and then removed. After a certain time period the signal is reapplied and more half cycles let through. Thus the ratio of *time on* to *time off* controls and determines the amount of power delivered to the circuit. This method of control is ideal for heating circuits.

3.8 Thyristor types

Thyristors are available with current ratings from one amp to several hundred amps. The larger ones tend to be of the stud type and resemble diodes but have three terminals. The smaller devices are similar to transistors, so if in doubt check the manufacturer's catalogue.

SUMMARY

Sections 3.5–3.8 have described in simple terms types of thyristor used in electronic circuits. The following important points should have been learned:

- That electricians must be able to recognize and identify the different types of thyristor;
- How thyristors are used in electronic circuits such as high speed switches circuits, dimmer circuits and motor speed control circuits. Students should also have gained a simple concept of their principle of operation;
- How to recognize BS 3939 symbols used for the different types of thyristor on circuit diagrams;
- The meaning of the expressions *phase control* and *burst control* in relation to thyristors;
- The advantages and disadvantages of using the thyristor in a.c. and d.c. circuits.

3.9 Triacs

A triac is another electronic switching device, but unlike the thyristor, which gives only half-wave control, the triac can be turned on during either the positive or the negative half cycles. Just one gate terminal provides this facility, and a further advantage is that the triac can be fired with either positive or negative gate signals. Like the thyristor, the triac will remain on until the current through the component falls below the holding current. Triacs do, however, require much higher voltage signals on the gate to turn them on compared to thyristors.

The triac is essentially two thyristors placed alongside each other, one inverted with respect to the other (inverse parallel). The three terminals of the component are called Main Terminal 1 (MT1), Main Terminal 2 (MT2) and gate (G). MT1 is the supply side, MT2 is the load side and G the gate.

Triacs are sometimes called *bidirectional thyristors* and are used in a.c. circuits for full-wave power control; however, whereas the thyristor is available with ratings of hundreds of amps, the triac is available only to about 15 A.

3.10 Diacs

These are two terminal devices containing bidirectional diodes that will not conduct until the voltage across them reaches a predetermined value known as the breakdown voltage.

Diacs are usually used as capacitor discharge devices. In this type of circuit the capacitor acts as an energy source of low impedance and, when it is discharged through the diac, provides a high energy pulse of short duration. Diacs are used extensively in triac and thyristor circuits to provide the pulse to the gate when the supply voltage waveform reaches the predetermined level (typically between 30 and 50 V). Since the device is bidirectional it triggers on both the positive and negative half cycles.

Circuits using these semiconductor devices will be discussed in Chapter 6.

SUMMARY

Sections 3.9 and 3.10 have described in simple terms types of triac and diac used in electronic circuits. The following important points should have been learned:

- That electricians must be able to recognize and identify the different types of triac and diac;
- How to use triacs and diacs in electronic circuits. The student should also have gained a simple concept of their principle of operation;
- How to recognize BS 3939 symbols used for diacs and triacs on circuit diagrams;
- The meaning of the expression *bidirectional thyristor* in relation to the triac;
- The advantages and disadvantages of the triac compared to the thyristor.

Test 3

Choose which of the four answers is the correct one.

(1) When testing a junction diode the correct results would be:

(a) high resistance in both directions,
(b) low resistance in both directions,
(c) high resistance one way, low resistance the other,
(d) no resistance in either direction.

(2) One major use for a zener diode is as a:

(a) high frequency diode,
(b) high power diode,
(c) current reference,
(d) voltage reference.

(3) The three terminations to a thyristor are called:

(a) cathode, anode and gate,
(b) input, common and output,
(c) base, emitter and collector,
(d) gate, source and drain.

(4) The thyristor is a:

(a) one layer semiconductor device,
(b) two layer semiconductor device,
(c) three layer semiconductor device,
(d) four layer semiconductor device.

(5) A triac can be turned on in:

(a) the positive half cycle only,
(b) the negative half cycle only,
(c) either the negative or positive half cycles,
(d) d.c. circuits only.

(6) The three connections on a Field Effect Transistor (FET) are:

(a) source, drain and gate,
(b) cathode anode and gate,
(c) emitter collector and base,
(d) T1, T2 and T3.

(7) When base current flows in a transistor:

(a) a much smaller current flows between emitter and collector,
(b) a much larger current flows between emitter and collector,
(c) much less current flows between source and drain,
(d) much more current flows between source and drain.

(8) A diac will not conduct until:

(a) the current in the circuit is high,
(b) the voltage across it reaches a predetermined level,
(c) the current in the circuit is low,
(d) it is connected in the correct polarity.

(9) Transistors are destroyed by:

(a) excess vibration,
(b) rough handling,
(c) excess internal heat,
(d) high switching speeds.

(10) A diode could be described as:

(a) a logic device,
(b) a polarity conscious switch,
(c) an interference suppressor,
(d) a field effect device.

Chapter 4
Integrated Circuits

4.1 Linear integrated circuits

Until 1960 electronic circuits were always made from a series of discrete components mounted on circuit boards and soldered into place. With further developments taking place in the use of semiconductor materials, manufacturers realized that it was possible to have more than one semiconductor device on the same piece of semiconductor material: integrated circuits (ICs) were thus born.

We have seen in our earlier work how semiconductor devices can be formed in the substrate material by a series of dopant diffusions; an integrated circuit is one that has many, sometimes even hundreds, of these regions. Some of the regions take the form of diodes and transistors, while in other areas resistors are formed by zig-zagging low resistance semiconductor material to increase the length of its resistance path. Some of the regions serve to isolate sections from each other by exploiting the properties of reverse bias junctions.

A number of identical ICs can be produced simultaneously on a disc of silicon. The disc is then cut into individual *chips* and connections brought out to pins. The chip is minute in size and it is necessary to mount it in some form of package so that it can be handled.

The dual in-line package

Modern ICs are usually housed in a black plastic package, rectangular in shape, with the connections of the integrated circuit being brought to pins, down each side of the package. The total number of pins depends on the complexity of the integrated circuit, but generally there are between 4 and 40. It is customary to have half the pins down one side and the other half down the other side. To enable the user to identify the pin number, one end of the IC package will be identified. Popular identification marks include:

(1) A dimple in the IC package next to pin 1;
(2) A half moon shape moulded in the package. This indicates the top of the IC, and pin 1 is on the left-hand side of this;
(3) A silver coloured band, again marks the top of the IC. Pin 1 is on the left-hand side as before.

No matter how many pins are used, ICs are always numbered like this. An

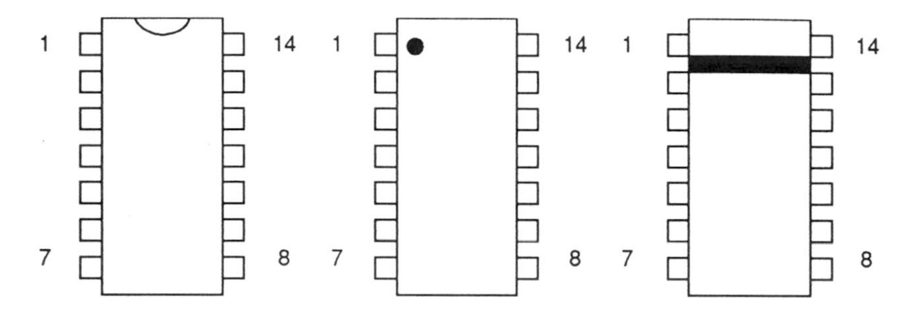

Fig. 4.1 Integrated circuit markings.

integrated circuit with 14 pins, for example, would have seven pins down one side and seven down the other. With the identification mark at the top and looking down on to the integrated circuit, pin 1 would be top left and pin 7 bottom left. At the same time pin 8 would be top right and pin 14 bottom right. An IC such as this would be described as a 14-pin dual in-line package, which is popularly abbreviated to 14-pin DIL. An example of this type of package is shown in Fig. 4.1.

The flat-pack package

Another style of packaging has recently appeared. This is square in shape and has the terminal pins around each side of the package. With this type of IC the pins are designed for soldering on top of the circuit board, as shown in Fig. 4.2. The main advantage of this form of package is that many more connections are possible as each side of the device provides pin-outs.

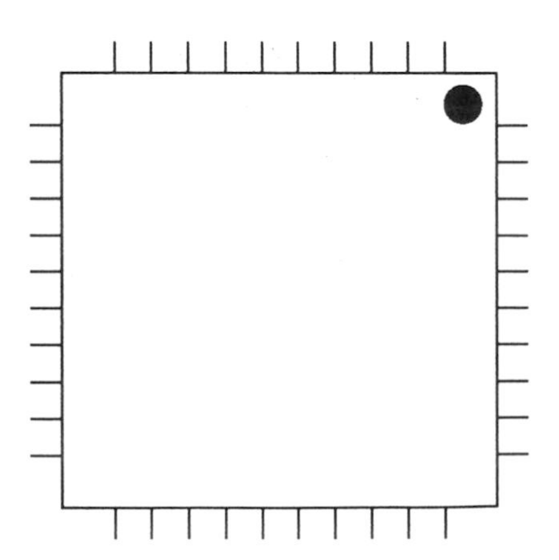

Fig. 4.2 Flat-pack IC.

Information sheet No.4.A.
Comparing op-amps

Type No.	Supply voltage	Power rating	Input resistance
uA741	± 5V to ± 18V	500mW	1.0 Mohm
LM308	± 5V to ± 18V	500mW	40 Mohm
NE531	± 5V to ± 22V	500mW	30 Mohm
CA3130E	5V to 16V	630mW	1.5 Tohm
CA3140E	4V to 36V	630mW	1.5 Tohm
LF351	± 5V to ± 18V	500mW	1.0 Tohm
TL071	± 2V to ± 18V	680mW	1.0 Tohm
TL081	± 2V to ± 18V	680mW	1.0 Tohm

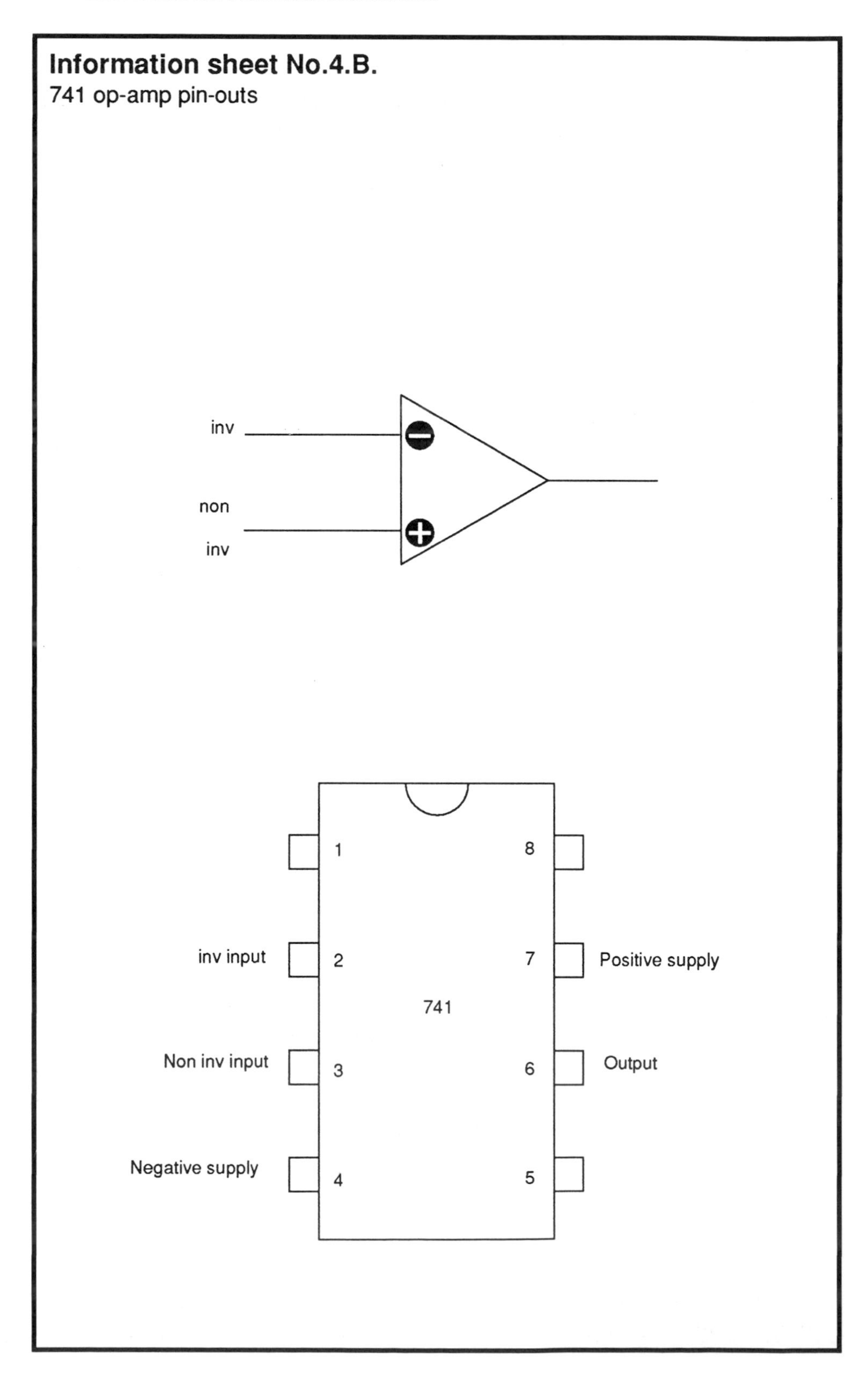
Information sheet No.4.B.
741 op-amp pin-outs
inv
non
inv
1
8
inv input
2
7
Positive supply
741
Non inv input
3
6
Output
Negative supply
4
5

4.2 Operational amplifiers

As the field of electronics widens and electronic circuitry becomes ever more complex, it is often difficult to remember how we ever got on without the use of certain integrated circuits. The operational amplifier (op-amp) is one such device.

It is a name given to an integrated circuit that historically found application in analogue computers, but it now has a much wider use. Perhaps the most popular op-amp in use today is the 741, but several others are in manufacture. Information Sheet No.4A gives details of op-amp parameters.

Several features are common to all op-amps:

(1) An inverting input;
(2) A non-inverting input;
(3) A high impedance (especially high with CMOS and FET types);
(4) A low input impedance;
(5) A large gain when operated without feedback, typically 100 000.

Usually having cight pins, thc op-amp has the ability to perform a wide range of basic electronic functions, depending on how other components are connected in circuit with it. Some functions include:

- Amplification of signals, both a.c. and d.c.;
- Generating waveforms;
- Integrating and differentiating waveforms;
- Amplifying the difference between two signals;
- Summing or subtracting of several signals;
- Some basic digital processes.

The circuit symbol and the physical layout of the pins are shown on Information Sheet No. 4B.

A look at the Information Sheet will show that pin 2 is marked with a '–' sign and pin 3 with a '+' sign. These two pins are not positive and negative connections to the op-amp, as you might at first imagine. In fact the '+' sign indicates the non-inverting input connection and the '–' sign indicates the inverting input connection. Although this may appear to be a little complicated at first, it simply means that if the input signal is applied to pin 2, not only will the output signal be amplified, but it will also be 180° out of phase with the input. If the input signal is connected to pin 3, the signal will be amplified, but no phase change will take place (see Fig. 4.3).

4.3 Digital integrated circuits

A digital integrated circuit operates in a non-linear mode. The inputs to an IC of this type respond to logical signals which are signals to which a truth value is

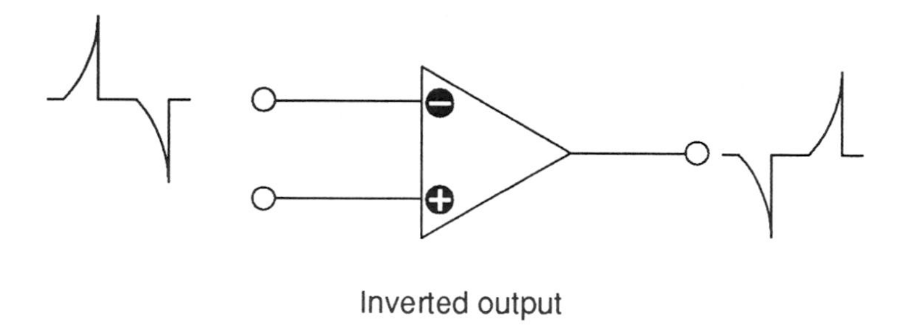

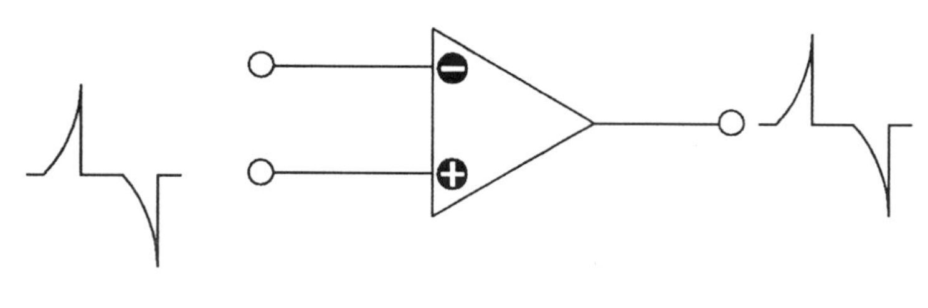

Fig. 4.3 Inverted and non-inverted signals.

attached. A true statement could be represented by a high voltage and a false statement could be represented by a low voltage. These are identified by the logical symbols 1 and 0 respectively.

Digital circuits employ two types of logic, sequential and combinational. Integrated circuits using sequential logic are those types of circuit with a memory capability and are beyond the scope of this book. Circuits built around logical signals and logic gates are called combinational logic gates.

4.4 Logic gates

A logic gate has one or more inputs; these are usually given the letters A and B for a two-input gate, or A, B and C for a three-input gate. A logic gate, however, has only one output and this is given the letter Q.

There are a number of different logic gates in use, and these are given what at first seem to be rather curious names; these are:

(1) AND
(2) OR

Information sheet No.4.C.
Circuit symbols for logic gates

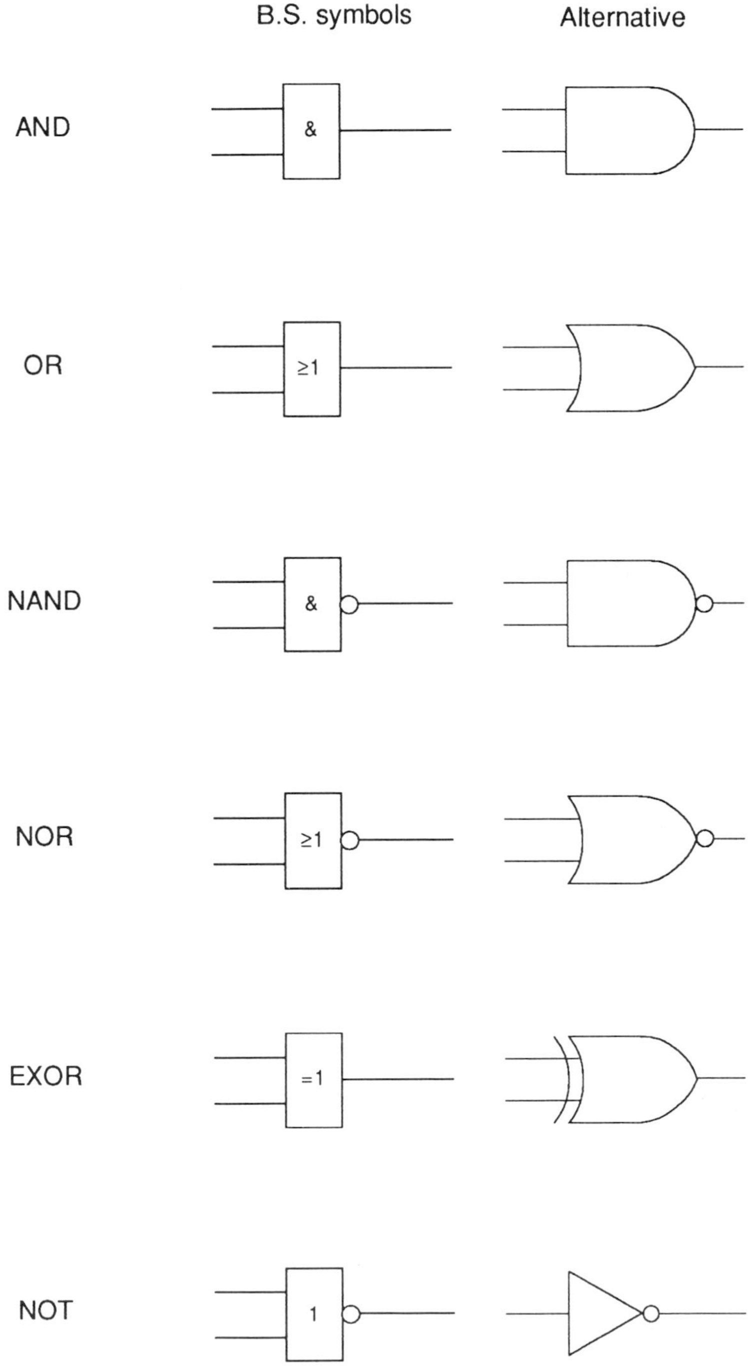

	B.S. symbols	Alternative
AND	&	
OR	≥1	
NAND	&	
NOR	≥1	
EXOR	=1	
NOT	1	

(3) NAND
(4) NOR
(5) NOT (sometimes called the inverter)
(6) Exclusive OR.

Each type has its own BS symbol, but circuit diagrams for foreign goods will use slightly different ones. Some of these symbols are shown on Information Sheet No. 4C.

Logic gates are designed to respond to digital signals only; analogue signals mean nothing to them. A door lock is a good example of a digital system; the door can be in one of two states, locked or unlocked. It could never be half locked (being realistic). A digital signal can have only two possible states. The signal can be low, a state represented by the figure 0, or high, represented by the figure 1. Analogue systems on the other hand have an infinite number of values at an infinite number of timing points (within reason). A car speedometer is a good example of an analogue device.

The truth table

Each type of logic gate will have its output at either 1 or 0, depending on the input conditions. The various permutations of these are usually displayed on a table which is known as a *truth table*. A truth table, therefore, displays the state of the output with respect to the input.

An analogy is often drawn between the action of the logic gates and a circuit using switches and lamps. It is possible to demonstrate the action of a logic gate in this way using switches to represent the input conditions and a lamp to indicate the output conditions. In an electrical equivalent circuit, when the lamp is lit this indicates logic 1, and when the lamp is out this indicates logic 0. These circuits are included purely as an aid to the understanding of particular gates, and should not be relied on when drawing truth tables. Let's look at some truth tables for devices with two inputs.

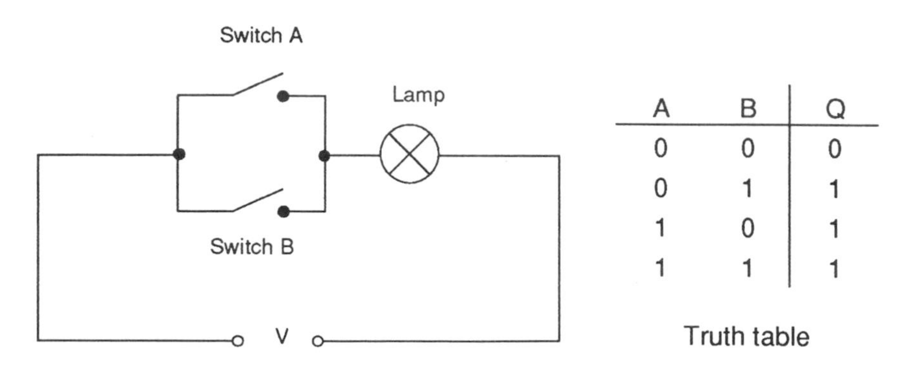

A	B	Q
0	0	0
0	1	1
1	0	1
1	1	1

Truth table

Fig. 4.4 The OR gate.

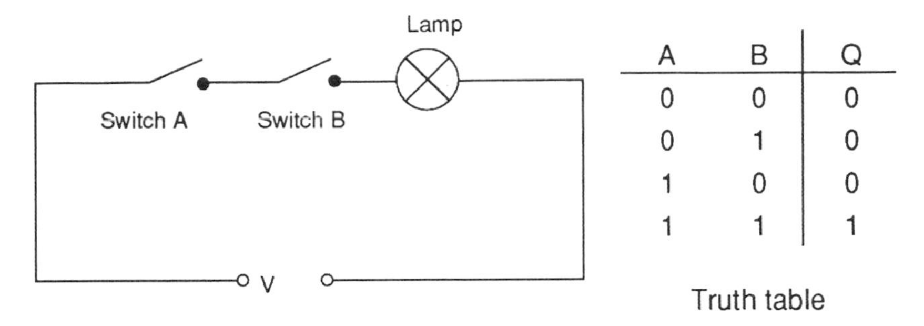

A	B	Q
0	0	0
0	1	0
1	0	0
1	1	1

Fig. 4.5 The AND gate.

The OR gate

The truth table for the OR gate together with its electrical equivalent circuit is shown in Fig. 4.4.

The first line of the truth table shows that if both inputs A and B are 0, then the output Q is also 0. If the input to either A or B is 1, then the output Q is 1. If both inputs A and B are 1, then the output Q will still be 1.

To summarize – if either A or B or both A and B are 1, then the output is 1.

The AND gate

The truth table for the AND gate together with its electrical equivalent circuit is shown in Fig. 4.5.

The first three lines of the truth table show that if any of the inputs to A or B are 0, then the output at Q will be 0. Only when both inputs are 1 will the output to Q be 1.

To summarize – the output will only be 1 if both A and B are 1.

Note: It is essential that students are able to work out the truth table for the OR gate and the AND gate for themselves. With this knowledge it is possible to derive the truth tables for the NOR and NAND gates, as we shall see later.

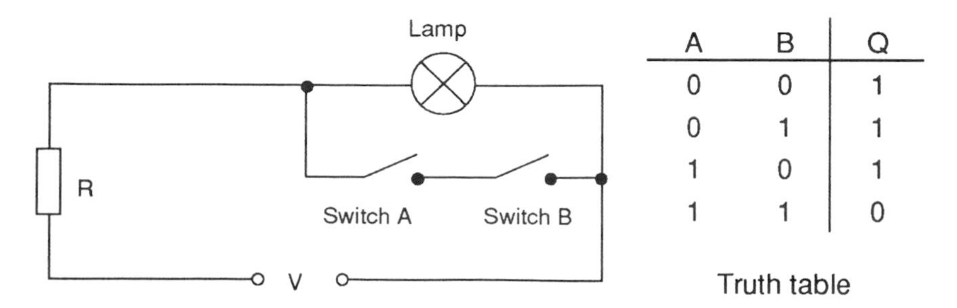

A	B	Q
0	0	1
0	1	1
1	0	1
1	1	0

Fig. 4.6 The NAND gate.

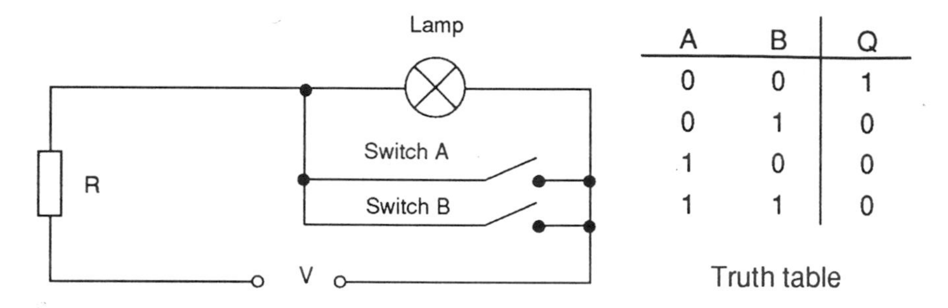

Fig. 4.7 The NOR gate.

The NAND gate

The truth table for the NAND (Not AND) gate together with its electrical equivalent circuit is shown in Fig. 4.6.

The truth table for the NAND gate shows that the output to Q is zero only when both the inputs to A and B are 1.

The NOR gate

The truth table for the NOR (Not OR) gate together with its electrical equivalent circuit is shown in Fig. 4.7.

The truth table for the NOR gate shows that if there is a 1 on the input to A or B then the output at Q will be zero. Only if both the inputs are 0 will the output to Q be 1.

To summarize – the output is 1 only if neither A NOR B NOR both are 1.

The NOT (or inverter) gate

The truth table for the NOT gate together with its electrical equivalent circuit is shown in Fig. 4.8.

Unlike the previous devices the NOT gate has only one input and one output;

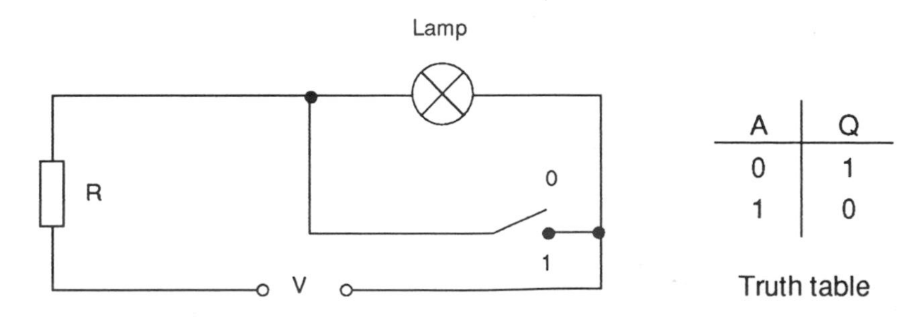

Fig. 4.8 The inverter gate.

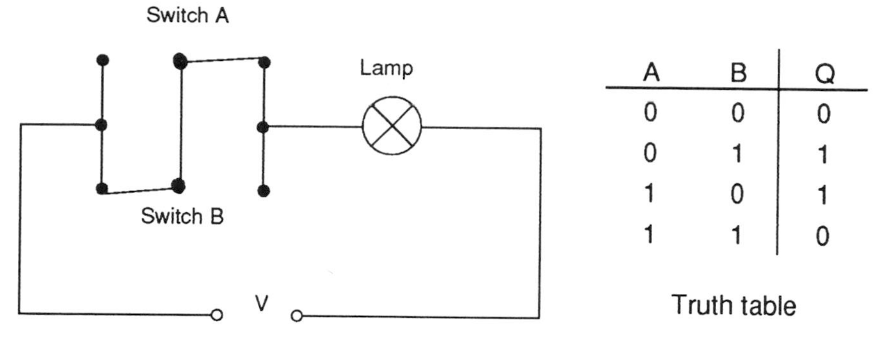

A	B	Q
0	0	0
0	1	1
1	0	1
1	1	0

Fig. 4.9 The Exclusive OR gate.

the truth table is therefore much simpler. The truth table shows that if the input at A is 0, then the output is 1. Conversely, if the input is 1, then the output is 0.

To summarize – the output at Q will be the opposite to the input at A.

The Exclusive OR gate

The truth table for the Exclusive OR gate together with its electrical equivalent circuit is shown in Fig. 4.9.

The Exclusive OR gate has a truth table very similar to the OR gate and varies only when both the inputs to A and B are 1.

The truth table shows that, like the OR gate, when the inputs are both 0, then the output is 0. When one of the inputs is 1, then the output is 1. Unlike the OR gate when both inputs are 1, the output is 0.

To summarize – the output is only 1 if the inputs to the gate are different.

Deriving NAND and NOR truth tables from AND and OR truth tables

As mentioned earlier, it is possible to derive the truth tables for the NAND and NOR gates from the AND and OR gates respectively. Look at the truth tables in

A	B	OR Q	NOR Q
0	0	0	1
0	1	1	0
1	0	1	0
1	1	1	0

Fig. 4.10 Deriving NOR from OR.

Fig. 4.10. The table shows the Q column for the OR gate and next to it the Q column for the NOR gate. Note that the values are opposite. Where there is a 0 in the OR gate there is a 1 in the NOR gate, and where there is a 1 in the OR gate there is a 0 in the NOR gate.

To summarize – to change a truth table for an OR gate into a truth table for a NOR gate, simply change the values in the Q columns to the opposite values.

Of course this technique applies equally well to the AND and NAND gates. The method of deriving the values for the NAND gate from those of the AND gate are exactly the same (see Fig. 4.11).

To summarize – to change a truth table for an AND gate into a truth table for a NAND gate, simply change the values in the Q columns to the opposite values.

		AND	NAND
A	B	Q	Q
0	0	0	1
0	1	0	1
1	0	0	1
1	1	1	0

Fig. 4.11 Deriving NAND from AND.

Logic gate technology

There are numerous methods of constructing logic gates, but we need not worry about the detail of the construction of logic gates, for it is the way that they behave that concerns us. Of the numerous methods of construction two types are common:

(1) The TTL (transistor-transistor logic). This has replaced the old RTL (resistor-transistor logic) and the DTL (diode-transistor logic) devices;
(2) The CMOS (pronounced see-moss) (complementary metal oxide semiconductor).

It is important at the design stage of a circuit using these devices that the designer decides whether to use TTL or CMOS devices as it is not recommended that the two types be mixed.

The TTL device

The TTL devices utilize the bipolar transistor in their construction. Perhaps the most important characteristic of this family is that they only require a 5 V d.c.

supply. In fact the supply must be very near 5 V (between 4.5 and 5.5) to guarantee satisfactory operation of the component. This, coupled with the fact that they draw relatively large currents from the supply (several mA per gate), rules them out for all but the very basic battery circuits.

The TTL devices do, however, switch very quickly, the operation of changing the output from 0 to 1 or 1 to 0 taking only 15 ns. They are not static sensitive and so require no special handling procedures and the LS type can handle frequencies in excess of 100 MHz.

The CMOS device

The CMOS devices utilize the field effect transistor (FET) and are the nearest thing so far to a perfect switch. The operating voltage is broad, from 3 to 18 V. Because of the FET construction, CMOS devices use exceedingly small amounts of power at low switching speeds – about one eighth the power of a similar TTL gate.

The biggest disadvantage of CMOS devices is that they are extremely static sensitive and require special handling procedures; these are detailed in Information Sheet No. 4D. Manufacturers do *buffer* some CMOS chips to try to protect them against static and these carry the letter B after their identification number, e.g. 4001B.

CMOS devices are a little more versatile than the TTL types and so they are more popular with circuit designers. They are much used in computer technology for registers, memories, etc.

Gate identification

It is usual for the logic gates to be produced in integrated circuit form. Each IC may hold several gates of one type. The description of a particular package and the gates within are described using a coded sentence.

The description of a particular IC might be: Quad 2 Input NAND gate. This expression would indicate that there are four NAND gates of the 2 Input type. Manufacturers give the IC's identification numbers: for a CMOS type IC it would be the 4011B type for the above gate, or the 4001B type for a Quad 2 Input NOR. All the CMOS devices start with a figure 4, and the letter B after the number signifies that it has a buffer fitted against static.

TTL devices are also readily identified, their identification numbers always starting with the figures 74 and the remaining numbers identifying what the IC does. For example, the 7400 is a Quad 2 Input NAND gate and the 7402 a Quad 2 Input NOR gate. You may see after the figures 74 the letters LS or ALS; this simply means that the device is a low power variety or a high speed type. For example, the 7408 is a Quad 2 Input AND gate and so is the 74LS08, only it is of the low power type.

Information sheet No.4.D.

Handling static sensitive devices

The very thin insulating layer between the gate and the substrate and its high resistance mean that field effect devices and also CMOS integrated circuits can be rendered useless by high voltages applied to its terminals, particularly the gate. To prevent damage to static sensitive devices there are a number of steps the engineer can take and a wide range of static discharge products on the market.

The engineer should:

(a) not wear clothes of synthetic material, or synthetic products in them.

(b) avoid wearing footwear with rubber or plastic soles.

(c) ensure that the soldering iron and other equipment is effectively earthed (which in the interests of safety should be anyway!).

(d) the engineer should discharge themselves to earth before handling these devices.

(e) keep all antistatic aids in place until the device is soldered or connected in circuit.

(f) use an antistatic workstation.

(g) sit on a chair made from conductive material.

(h) use conductive transit boxes when transporting static sensitive devices. Alternatively foam wrapped in aluminium foil makes a very good substitute.

Information sheet No.4.E.

Comparing logic gate parameters

Logic series	Fan-in	Fan-out	Propagation delay (ns)	Supply voltage	Power/gate (mW)
7400	10	10	10	5	22
74H	10	10	6	5	22.5
74L	10	20	33	5	1
74S	10	10	3	5	20
74LS	10	50	9	5	2
74ALS	10	20	4	5	1
4000	10	< 50	30	3 to 18	1

Fan out and fan in

An explanation of these terms may be required if we are to understand fully the information on integrated circuit logic gates given on Information Sheet No. 4E.

Logic gates are connected together in order to get the required logic effect that we want. The expression *fan out* refers to the maximum number of logic gates that may be connected to the output of a logic gate, and still guarantee that it will function correctly. CMOS gates have *fan out* properties superior to those of the TTL device.

The number of logic gates that can be connected to the input of a logic gate and still guarantee that it will function satisfactorily is called *fan in*. Both CMOS and TTL gates have equal *fan in* properties.

SUMMARY

This chapter has described in simple terms types of integrated circuit used in electronic circuits. The following important points should have been learned:

- That electricians must be able to recognize and identify the different types of packaging used to house integrated circuits;
- The concept of integrated circuits; and how to use them in electronic circuits;
- How to recognize BS 3939 symbols used for integrated circuits on circuit diagrams; and how the pins of integrated circuits are identified and numbered;
- The meaning of the expressions *linear integrated circuit* and *digital integrated circuit*;
- How to recognize an operational amplifier (how it is utilized in electronic circuits should also be understood, in simple terms);
- That digital signals have two states – 0 (low) and 1 (high) – and that the *truth table* displays the state of the output with respect to the input of a gate;
- The meaning of the abbreviations TTL and CMOS and the advantages and disadvantages of these – particularly the precautions to be taken when handling the CMOS device;
- The meaning of the expressions: *fan in* and *fan out*.

Test 4

Choose which one of the four answers is the correct one.

(1) Special handling procedures should be used when working with:

(a) TTL circuits,
(b) operational amplifiers,
(c) CMOS devices,
(d) analogue devices.

(2) Two switches in series have the same logic as a two input:

(a) AND gate,
(b) Exclusive OR gate,
(c) OR gate,
(d) NAND gate.

(3) Digital signals have:

(a) only one level,
(b) only two levels,
(c) a sawtooth waveform,
(d) a sinusoidal waveform.

(4) TTL circuits require an accurate supply voltage of:

(a) 5 V,
(b) 7.5 V,
(c) 1.8 V,
(d) 2 V.

(5) The '+' sign on pin 2 of an op-amp circuit symbol indicates:

(a) the non-inverting input,
(b) the inverting input,
(c) that it is a polarised device,
(d) this is the positive terminal.

(6) A dimple on one end of a DIL packaged IC enables the user to:

(a) fit an insertion tool,
(b) line up the integrated circuit,
(c) identify the pin numbers,
(d) know whether it is buffered or not.

(7) A truth table displays:

(a) whether the gate is CMOS or TTL,
(b) the maximum operating frequency of logic gates,
(c) the state of the output with respect to the input,
(d) how the circuit works in reality.

(8) Which type of gate has a superior fan-out:

(a) TTL,
(b) CMOS,
(c) analogue,
(d) digital.

(9) Which type of gate often referred to has the inverter?

(a) NAND,
(b) OR,
(c) AND,
(d) NOT.

(10) An IC described as a quad two input NAND gate will have:

(a) one logic gate,
(b) two logic gates,
(c) three logic gates,
(d) four logic gates.

Chapter 5
Electronic Circuit Assembly

5.1 Types of diagram

Before any work is carried out on the assembly of electronic circuits, it is first necessary to have available some type of drawing showing how all the individual components are connected together. There are a number of different types of drawing which can help us in different ways and it might be a good idea to look at some of these.

Circuit diagrams

These are the diagrams that readers will be most acquainted with. A circuit diagram shows the interconnections of components in their simplest form. It displays useful information regarding how each section of the overall circuit functions, and in addition it has some details on the values and types of components. It will perhaps indicate voltage levels at certain important points and the voltage waveform may be given. Circuit diagrams should be drawn using BS 3939 symbols, so these must be learnt if the diagrams are to be understood. A typical circuit diagram is shown on Information Sheet No. 5A.

Block diagrams

Block diagrams serve to give an overview of how the complete system works, with little reference to components. The function of each particular stage of the circuit is represented by a block; the blocks are then joined together by lines. Arrows are used to indicate directions of input and output signals and logic circuits use block diagrams extensively. A typical block diagram is shown in Fig. 5.1.

Layout diagrams

While circuit diagrams show the theoretical position of components, layout diagrams show pictorially how the components in the circuit diagram are actually laid out. It may show the circuit board and give some idea of the physical size of the components. The layout diagram, therefore, goes hand-in-hand with the

Information sheet No. 5.A.

Typical circuit diagram

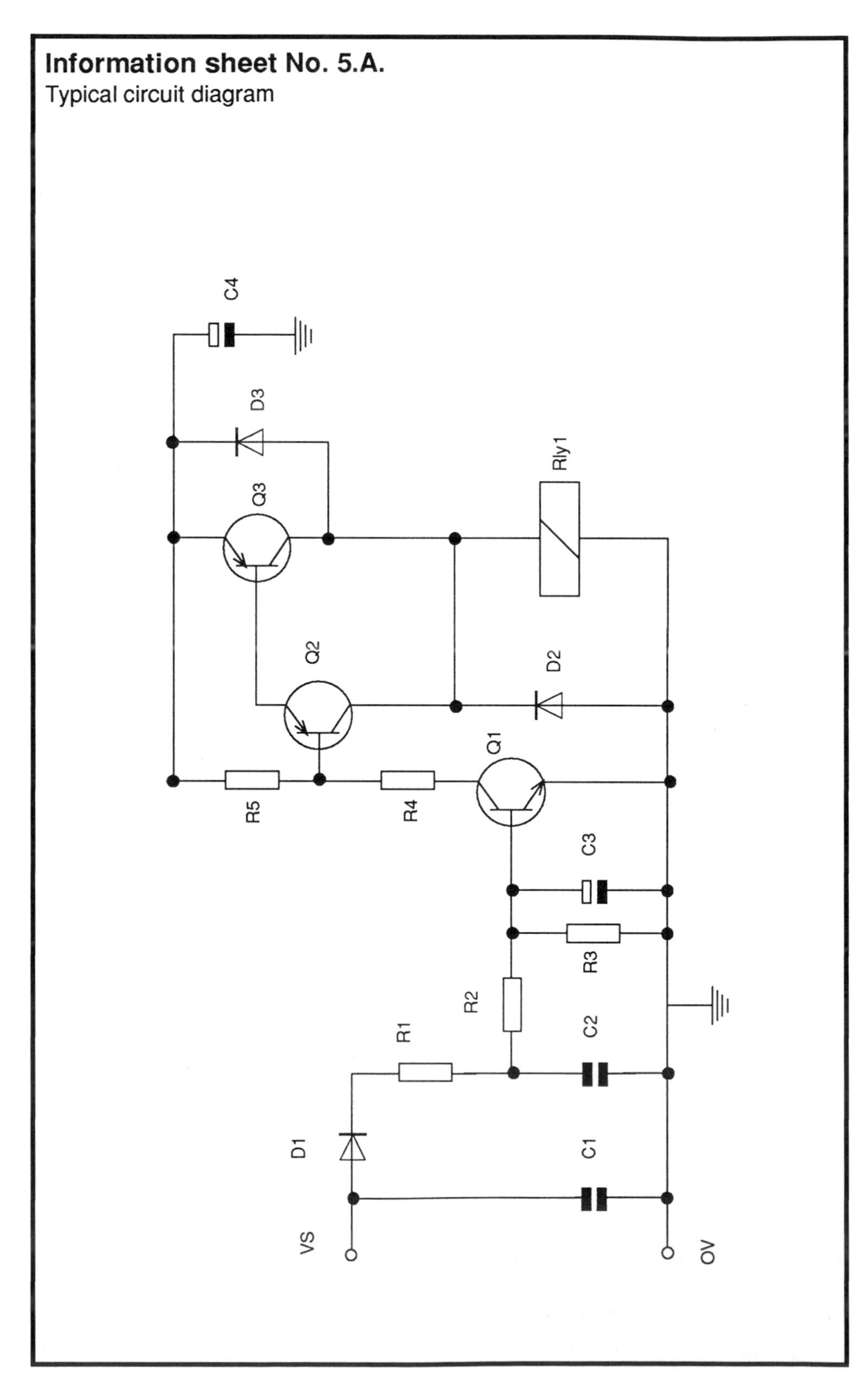

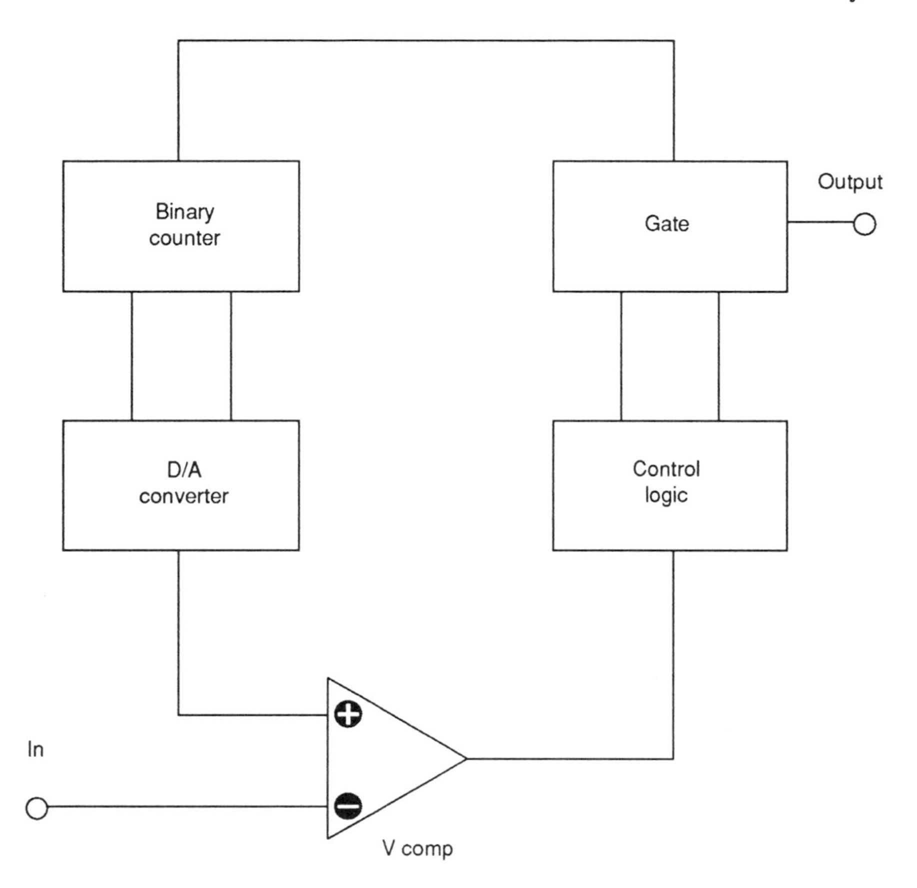

Fig. 5.1 Typical block diagram.

circuit diagram, and together they simplify component recognition. A typical layout diagram is shown on Information Sheet No. 5B.

Wiring diagrams

These are generally used to indicate the routes for external connecting wires, e.g. switches, indicating lights, DIN sockets. If the cables are identified by a colour this may be included on the wiring diagram where BL = blue, R = red and BlGr = blue with a green tracer. This is particularly popular on diagrams used for the wiring of automobiles and hi-fi equipment. A typical wiring diagram is shown on Information Sheet No. 5C.

Positional reference systems

If we wanted to know where Oxford Street was in London, we would consult an *A to Z Guide*. The index would refer us to page 55, then row 3, line 7. Turning to

Information sheet No.5.B.

Typical layout diagram

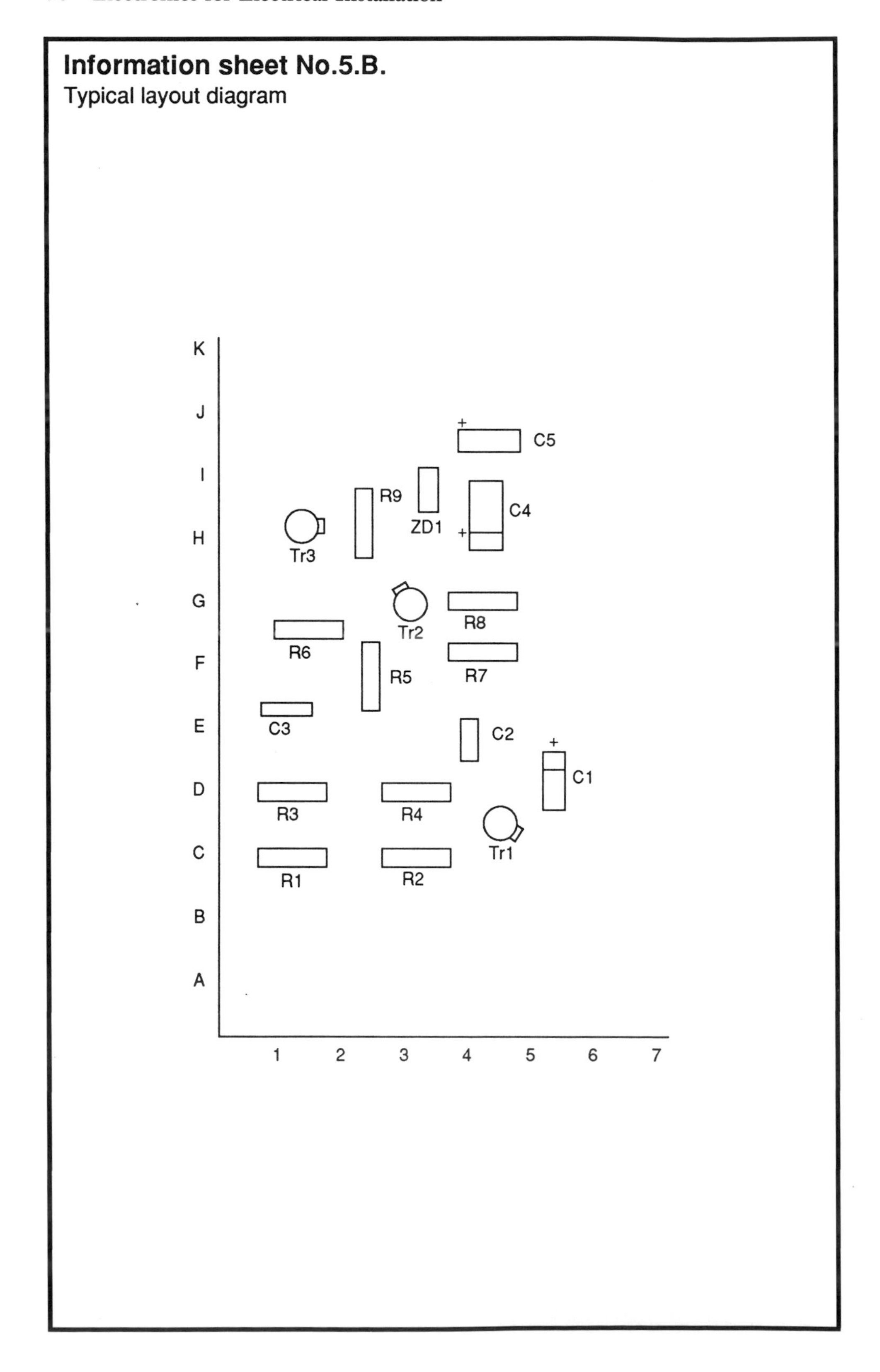

Information sheet No.5.C.

Typical wiring diagram

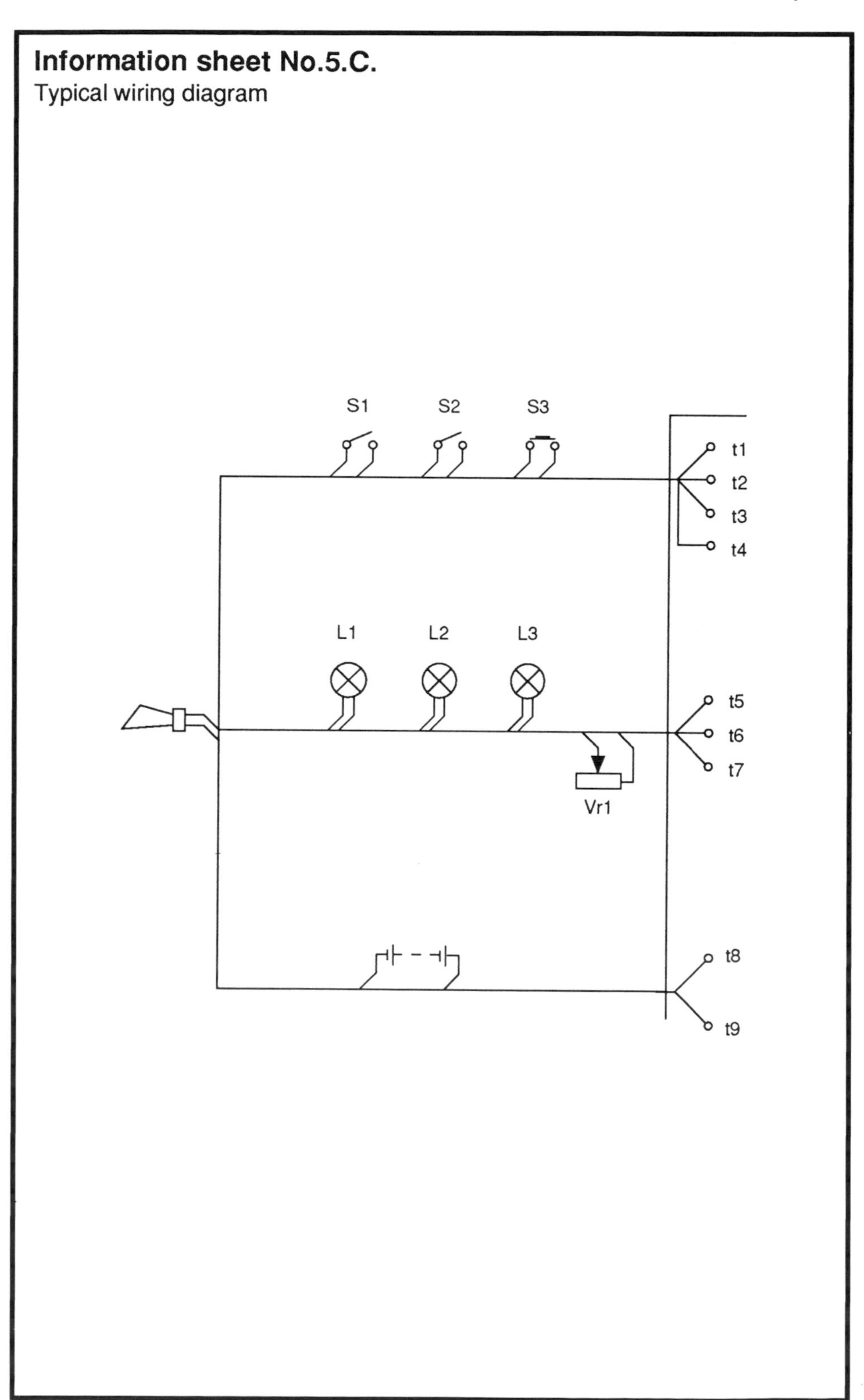

Information sheet No.5.D.

Component positional reference system

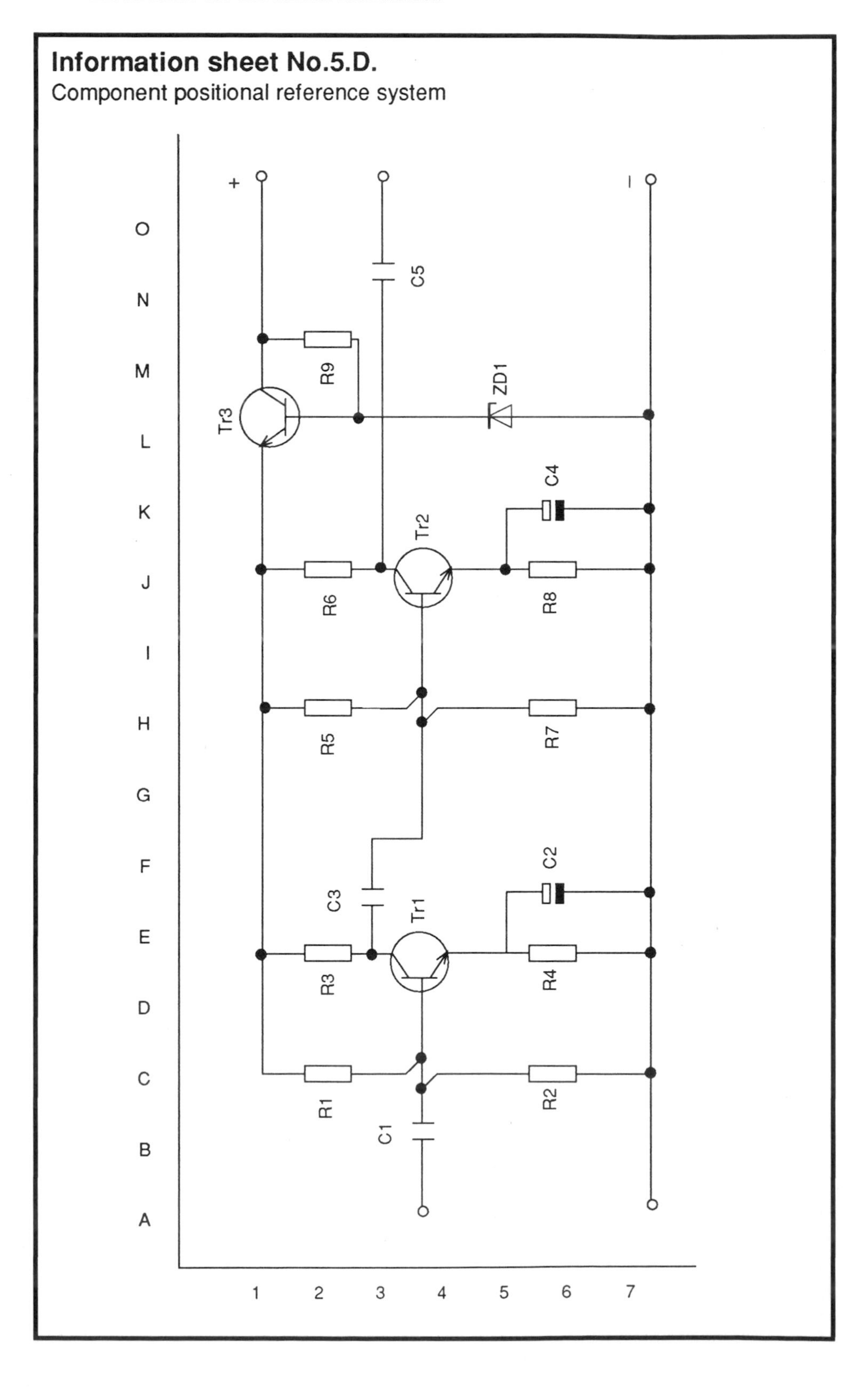

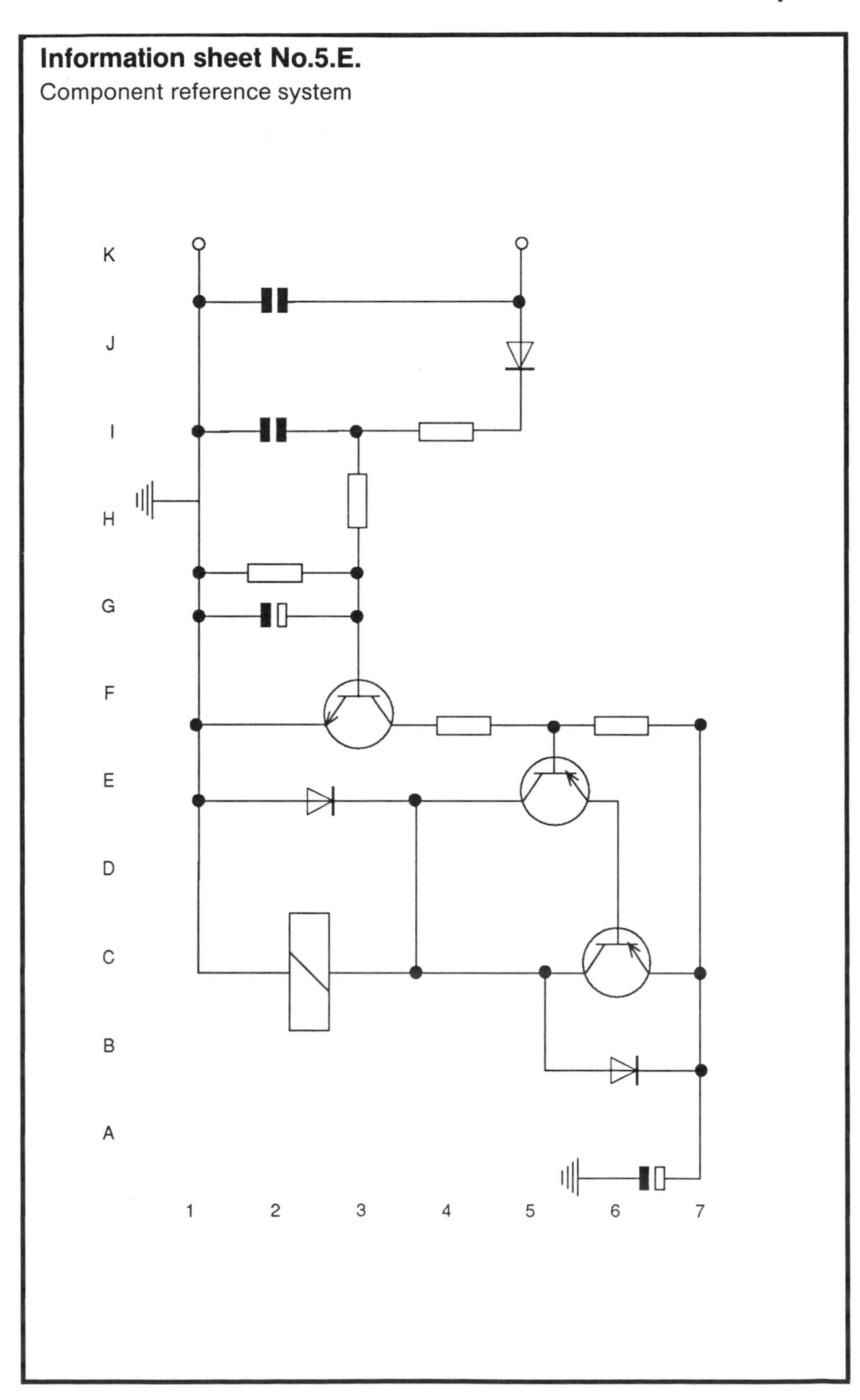
Information sheet No.5.E.
Component reference system
K
J
I
H
G
F
E
D
C
B
A
1
2
3
4
5
6
7

page 55 and looking at the intersection of row 3 and line 7, we find Oxford Street. This is an example of a street reference system. Component reference systems are identical.

An example of this can be found using the positional reference system shown on Information Sheet No. 5D. If, for example, we wish to find the location of a particular resistor, R3, we would first look up the index of components, which would refer us to a particular location, say E2. Component R3 would be found where row E and line 2 intersect. Information Sheet No. 5E shows the same circuit on a layout drawing; it shows how the positional reference system can be used on these drawings too.

5.2 Soldering

It will probably come as no surprise to read that connecting electronic components in circuit by the use of the soldered joint is still the most widely used of all connection methods.

A correctly soldered joint has good mechanical properties, low electrical resistance and high conductivity, and it is a quick and easy way of connecting components and wires. Moreover, should the component require to be removed from the circuit, perhaps because of failure, then, once the skills have been learnt it is a relatively easy task to desolder the component.

Solder for electronic work is 60% tin and 40% lead. With this ratio the melting point is around 188°C. The transition period for the molten solder to return to a solid state is quite short, and there is a short period during which the solder is semi-liquid. Solder with a ratio of 60/40 is sometimes referred to as *Tinman's solder*; as it is used by tinsmiths, it has a good wetting/flow capability and is ideal for our purpose.

Solder used in electronics is supplied on reels and looks rather like wire; it has one or more cores of non-corrosive flux running through its centre. The role of the flux is to keep the parts to be soldered clean by dissolving the oxide film that forms, and to prevent the heat of the iron and the oxygen from the air from causing further oxidization. Solder comes in several different diameters for different sizes of component.

Electric soldering irons are rated by their wattage, and range from 12 W up to a massive 125 W, 25 W being about right for repair work. The tip of the soldering iron is called the *bit* and has a temperature of between 210°C and 390°C. Irons can be pencil-shaped or be in the form of a gun. Even gas-heated ones are available. The bits come in various shapes; some of the more common ones are shown in Fig. 5.2.

The most important thing to remember when commencing any form of soldering is cleanliness. Make sure the areas to be soldered are clean and free of grease or oil. Circuit boards should always be stored in their packaging in a warm dry atmosphere and components kept in those little plastic drawers to keep them clean and free from harm. Keep your fingers off the parts to be soldered, because

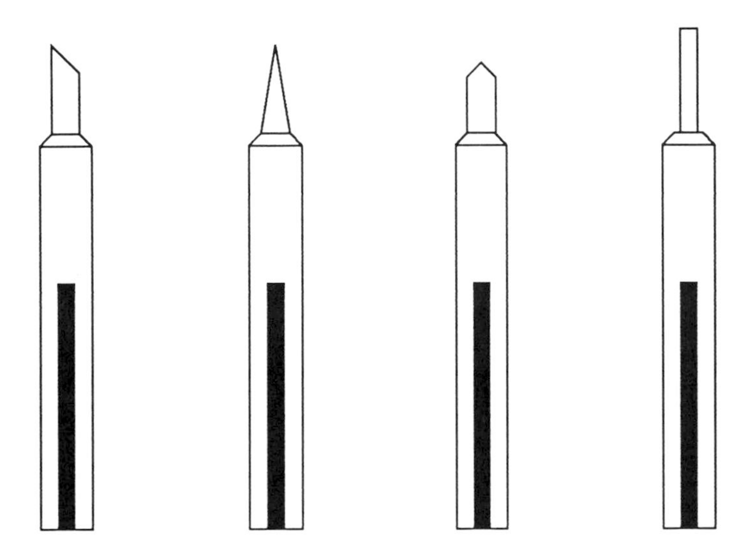

Fig. 5.2 Soldering bits.

no matter how well you wash your hands they will still retain body oils.

Before commencing soldering, the bit of the soldering iron should be *tinned*, i.e. a thin layer of solder should cover the tip of the iron. Do this by first removing any dirt or grease from the bit, either by rubbing with emery paper in the case of copper bits, or wiping with a cloth or damp sponge whilst the iron is hot, for iron- or chromium-plated bits. Next touch the heated bit with multi-core solder; let it spread, then wipe off any surplus solder with a cloth or damp sponge. Keep the soldering iron in a stand whilst not in use; this will not only keep the tip clean, but will reduce the chance of accidents.

It is also a good idea to *tin* any wire leads of the components you are going to use if this is possible. The idea of tinning is not to provide solder for the actual joint, but to provide good heat contact between the bit and the parts to be soldered.

Place the component in position; bending the component lead over slightly can help to keep it in place; do not overdo this because it tends to make desoldering more difficult. In order to pre-heat the work, apply the heat to both the component lead and the copper track of the circuit board simultaneously with the tip of the iron. The heating will take only two or three seconds and then the solder can be applied to the joint. When the solder has run around the joint, take the heat away as quickly as possible. Some semiconductors can be damaged by excessive heat; a further precaution you can take while soldering is to clamp your narrow-nosed pliers on the components lead, between component and joint. This will cause the pliers to act as a heat shunt, absorbing any excess heat; a crocodile clip will achieve the same result. The joint should be allowed to cool without being disturbed and should have a bright finish. If the joint looks dull, then too much time was spent on the job and you have lost most of the tin from the joint; this will reduce its mechanical strength.

The **wrong** way to solder is to load the bit with solder, and then transfer this

Information sheet No.5.F.

Steps for a correct soldered joint

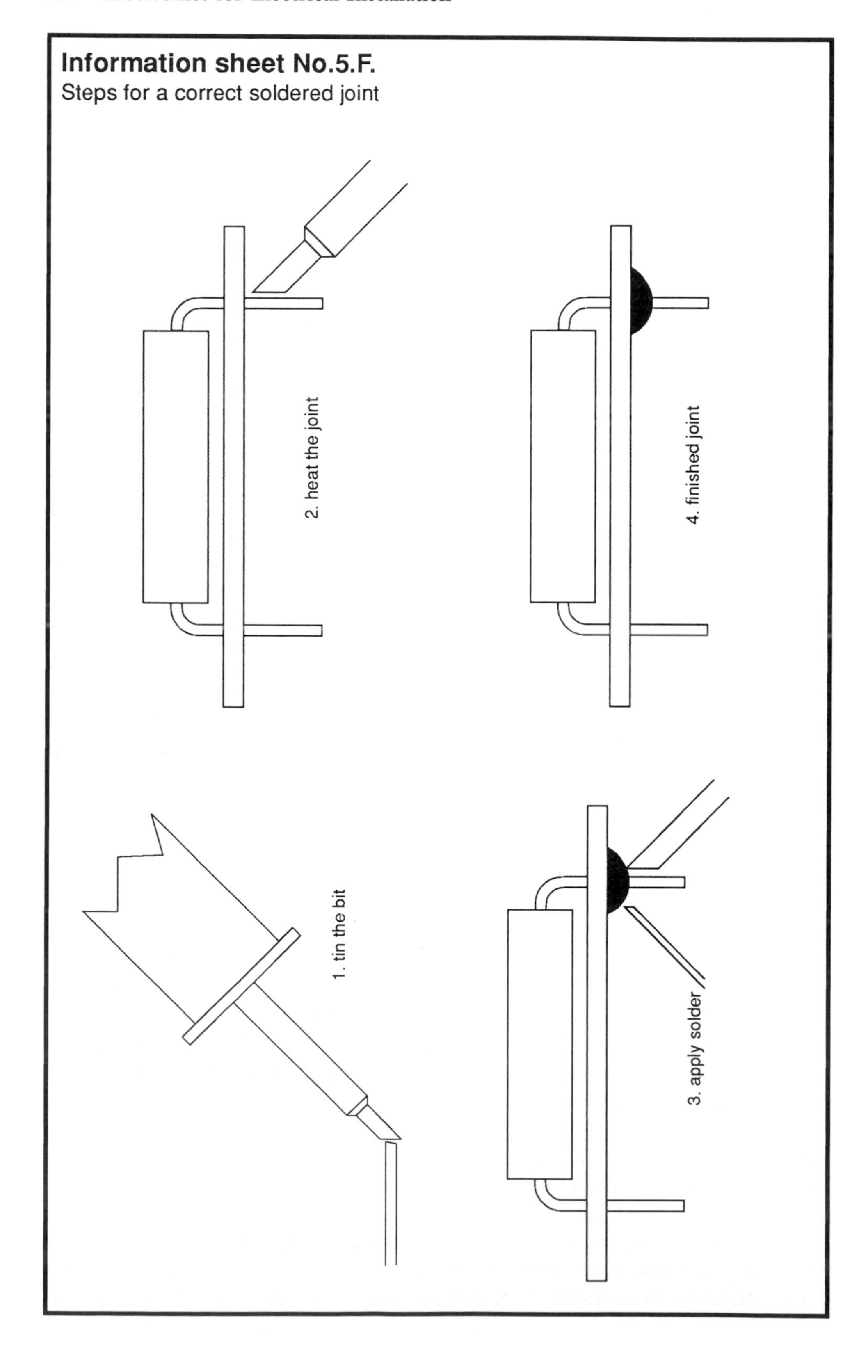

to the joint. This results in what is known as a *dry joint*, the useful properties of the flux being exhausted before the joint is made. In addition, the temperature of the solder will be too low and simply mould itself around the lead rather than being soldered to it. Dry joints have high electrical resistance and should be avoided at all costs. The correct steps for soldering are shown on Information Sheet No. 5F.

5.3 Desoldering components

Two methods of desoldering are in common use:

(1) Desoldering braid;
(2) The desoldering suction tool.

Desoldering braid is used by offering the braid to the soldered joint at the same time as the hot soldering iron. The braid heats up and the solder melts. The braid is manipulated so that it soaks up the molten solder.

Desoldering suction tools come in several shapes and sizes, from the simple 12 V d.c. portable types suitable for field work, to the large desoldering station complete with fume extraction equipment. The more sophisticated ones have cool running heads, footpedal switches, choice of pressure or vacuum modes, antistatic heads and RCD protection. More convenient to use, the suction tool has the further advantage that the hot solder is sucked up into the tool.

A popular model works by pressing down a spring-loaded piston to prime it. The solder is then melted around the joint to be desoldered. On pressing a button at the side of the tool, the piston is released and the solder is sucked up into the tool. Waste solder will be ejected after cooling, when the tool is next primed.

5.4 Wire wrapping

This long-standing technique is as good (some would argue better) for certain applications as the soldered joint. It was first introduced by one of the big telephone companies in the USA to meet a need for a fast, cost-effective method of terminating the hundreds of connections in their circuitry.

There are a number of different tools on the market; these range from hand tools suitable for a small number of terminations, to the spool-fed electrically operated tools for mass production. The tools use two methods of terminating:

(1) The non-stripping method;
(2) The stripping method.

In the *non-stripping method* square pins are inserted into the circuit board. The electrically operated tool wraps the wire round the pin tightly and in so doing causes the sharp corners of the pin to cut through the insulation, effecting a good

Information sheet No.5.G.

Correct and incorrect wire wrap terminations

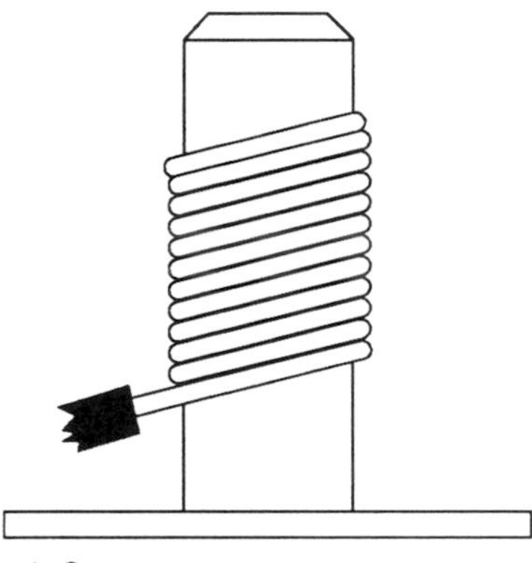

a) Correct wrap

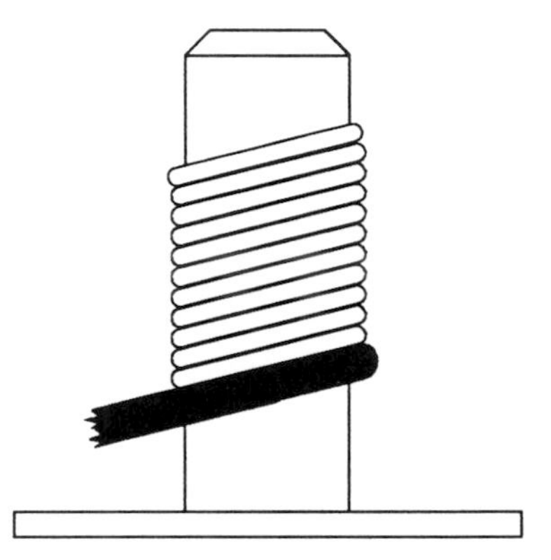

b) Modified wrap (insulation taken round terminal)

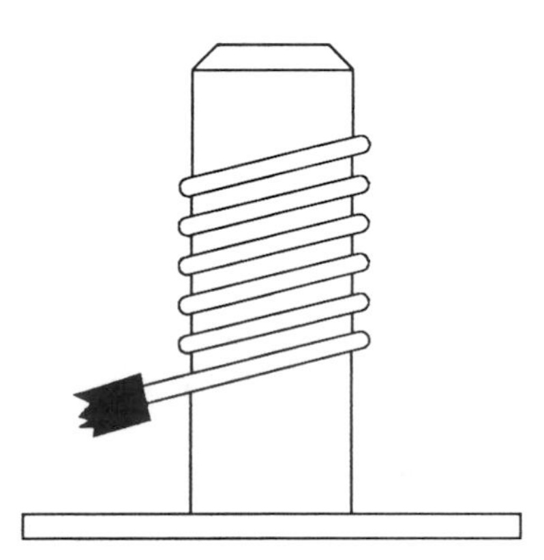

c) Incorrect wrap spaces between turns

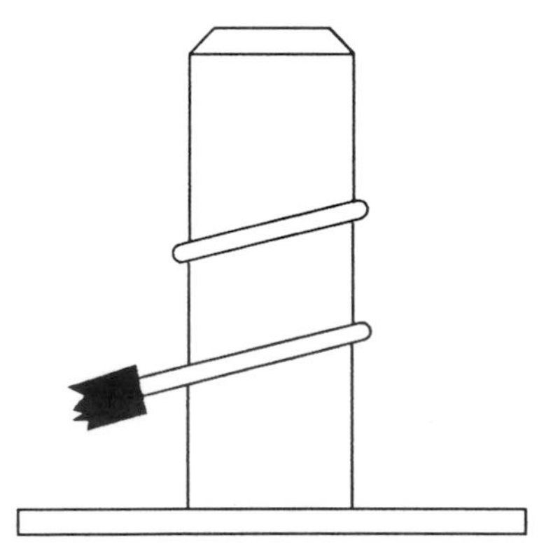

d) Incorrect wrap insufficient turns

connection. To avoid overwrapping of the small diameter cables, the tool, which spins at around 4000 RPM, has a built in *back force* device.

The *stripping method*, as its name implies, requires that the wire is stripped. The more expensive machines have *cut-wrap-bits* which automatically and simultaneously strip the cable and terminate it. Some of these machines produce a modified wrap; the first wrap includes the insulation-affording protection against vibration.

The wires come in three popular sizes: 24 AWG (American Wire Gauge), 0.40 mm; 26 AWG, 0.50 mm; and 30 AWG, 0.25 mm. To avoid a poor termination, it is essential that the sizes of the bit, pin and wire are the same.

Removing the wire from the terminal is almost as easy. The tool is placed over the terminal and rotated in an anti-clockwise direction until the wrap is undone. Most tools will carry out both wrapping and unwrapping operations (see Information Sheet No. 5G).

5.5 Circuit boards

The types of circuit board available for circuit assembly are almost as numerous now as the circuits designed to go on them. Each of the numerous designs falls into one of three broad categories:

(1) Printed circuit boards (PCBs);
(2) Strip boards;
(3) Matrix boards.

Printed circuit boards (PCBs)

We learned earlier that mass production techniques had brought down the price of PCBs dramatically. The process involves photo-resist boards, transfers, drafting sheets, ultraviolet exposure units, heated processing tanks, PCB bubble etch tanks, water deionizers, spray wash tanks, drying racks, etc. The process has been semi-automated for the production of repeat units and it can be seen from this that production at this level is an intricate business.

The *one off* board can still be produced but this should be done under the supervision of a competent person and all appropriate protective clothing should be worn and safety rules observed.

The idea is that the required circuit is drawn out on a copper-clad epoxy glass circuit board with an etch-resist pen. Allow about 15 min for the ink to dry and then immerse the prepared board in a bath of etchant solution such as ferric chloride. After etching, it will be seen that all surplus copper which had not been protected will have gone, leaving the desired circuit intact. The board is washed and the etch-resist ink removed with a PCB eraser/cleaner and placed on a rack to drain and dry.

When the board is thoroughly dry, holes to take the components are carefully

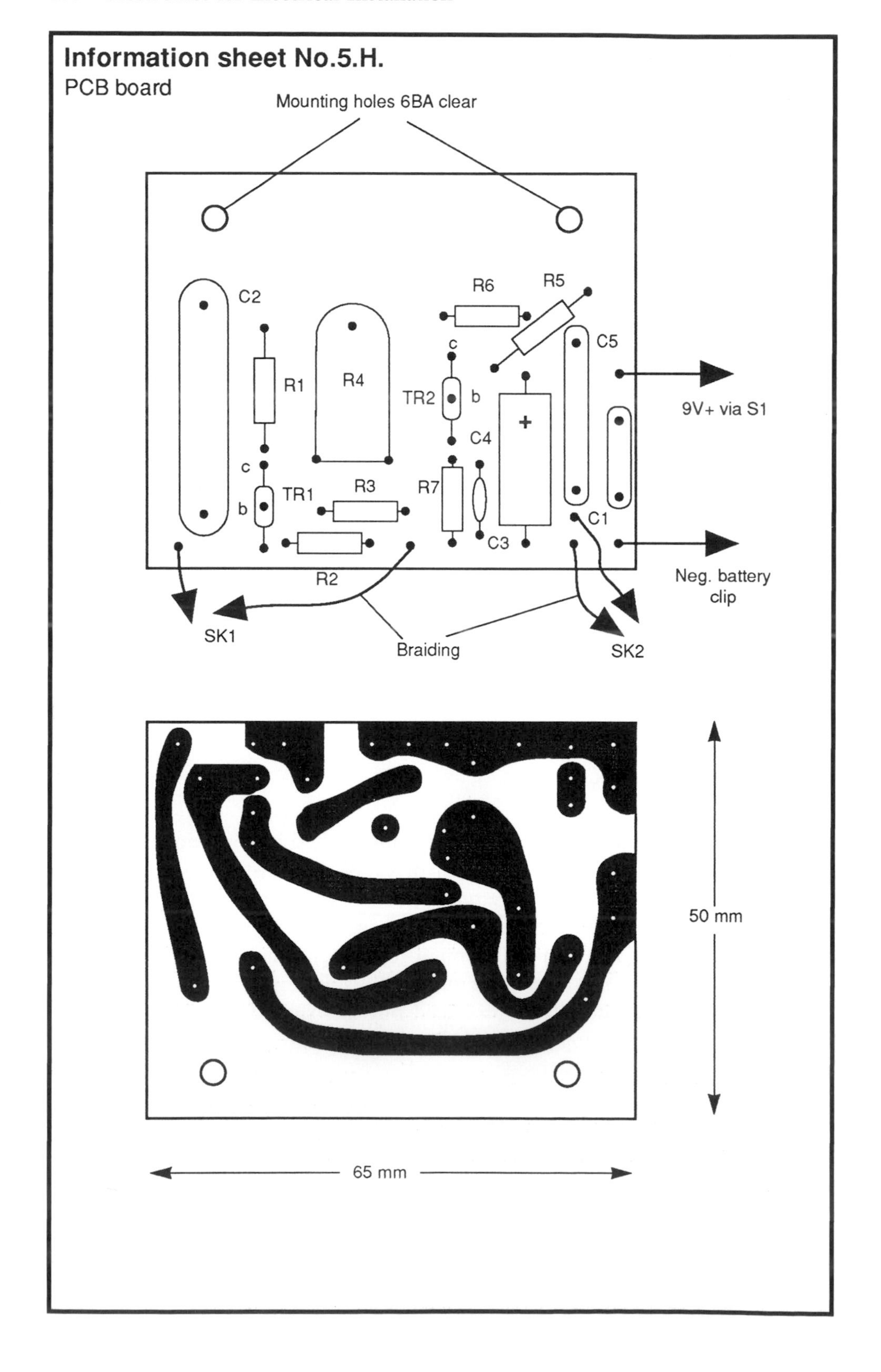

Information sheet No.5.H.
PCB board
Mounting holes 6BA clear
C2
R6
R5
C5
R1
R4
c
TR2
b
9V+ via S1
+
C4
c
TR1
R3
R7
b
C1
C3
Neg. battery clip
R2
SK1
Braiding
SK2
50 mm
65 mm

Information sheet No.5.I.

Strip board

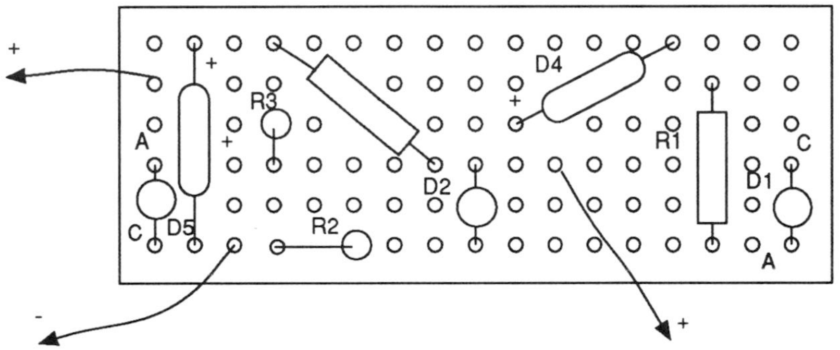

+
+
D4
R3
+
A
+
R1
C
D2
D1
C
D5
R2
A
-
+

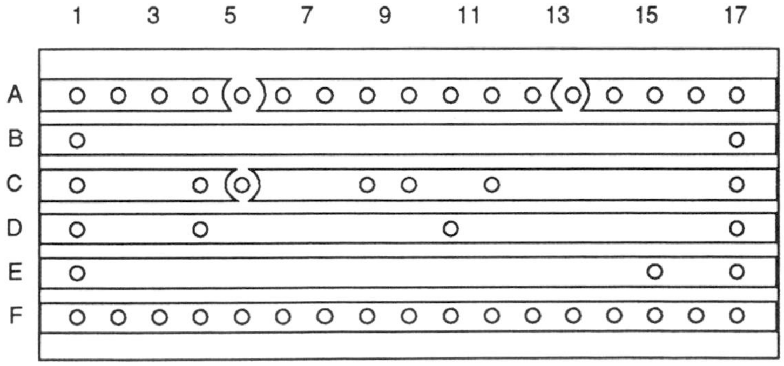

1 3 5 7 9 11 13 15 17

A
B
C
D
E
F

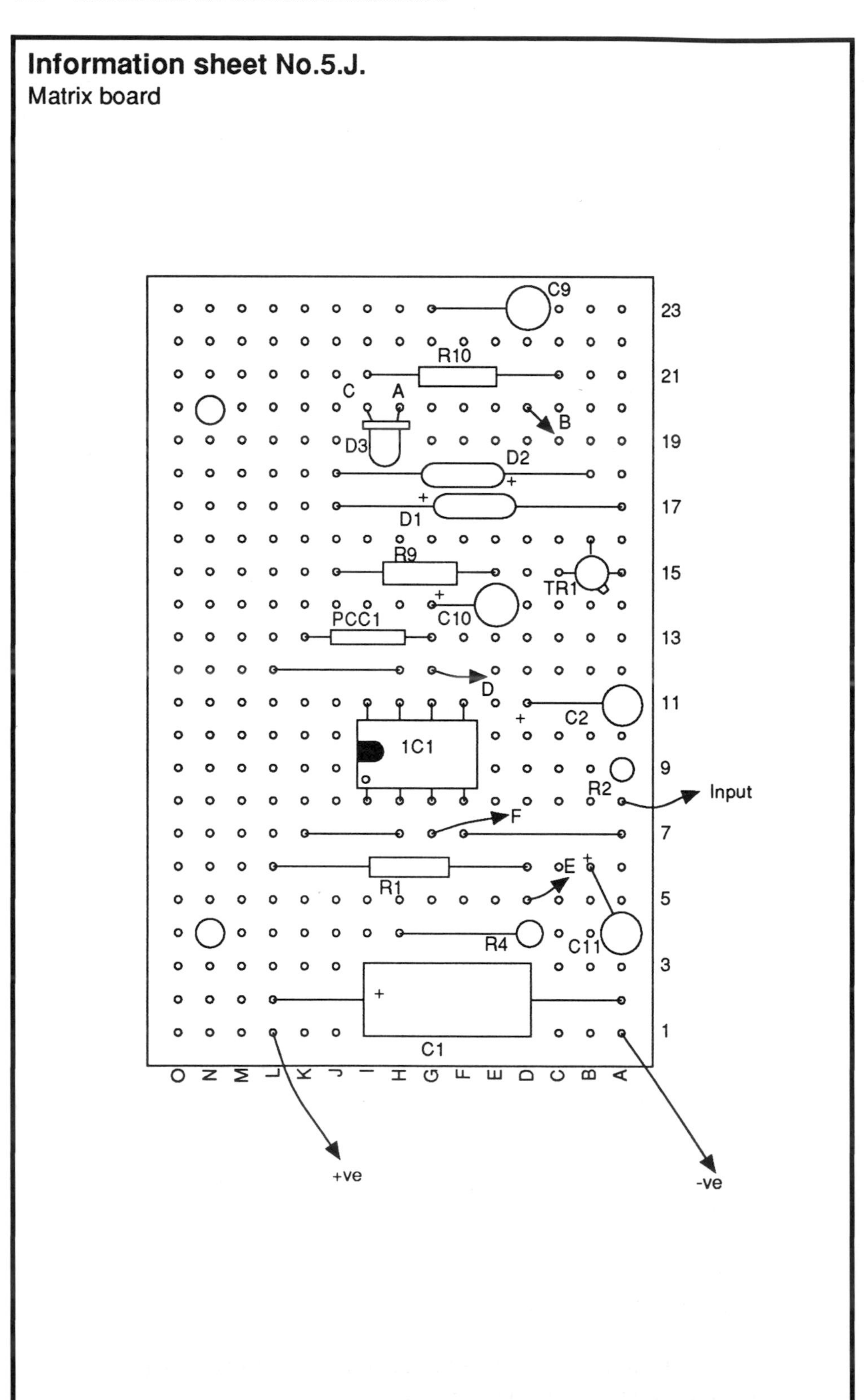
Information sheet No.5.J.
Matrix board
C9
R10
C
A
B
D3
D2
D1
R9
TR1
PCC1
C10
D
C2
1C1
R2
Input
F
E
R1
R4
C11
C1
23
21
19
17
15
13
11
9
7
5
3
1
O N M L K J I H G F E D C B A
+ve
-ve

drilled in the remaining copper sections. The components are then inserted from the plain side of the board and soldered to the copper strip. The copper is attached to the board by adhesive, and excessive heat or rough handling will result in this peeling from the board; therefore due care and attention should be taken. See Information Sheet No. 5H for details.

Strip boards

These consist of rows of copper tracks bonded to a synthetic resin bonded paper board (SRPB). The tracks are pierced at 0.1 inch centres to accept terminal pins and provide a simple form of printed circuit with the copper strips forming the interconnections between components. The tracks can be cut where necessary by inserting a simple cutting tool into a hole where the break is required, a few clockwise turns producing a neat break. The same effect can be obtained by use of an appropriately sized high speed twist drill turned by hand until the copper is severed. While this avoids the extensive preparation required for the PCB method mentioned above, it does require some careful rearranging of the circuit diagram for the most effective use of the board to be obtained.

The components are placed in position and soldered in place as described above, one big advantage being that many components these days have their connection leads arranged to suit the 0.1 inch pattern of holes. See Information Sheet No. 5I for details.

Matrix boards

These consist of synthetic resin bonded paper boards (SRPBs) pierced with holes of 0.052 inch diameter on a 0.1 inch grid. They are designed to accept press-fit single-sided and double-sided terminal pins. The circuit is laid out as near as possible to the circuit diagram to avoid mistakes. Pins (preferably of the double-sided type) are pushed into the board at the points where connection between the component and the circuit wiring is to take place. The components are then soldered to the top side of the board in the case of the double-sided pins, and interconnecting wires soldered to the underside of the board. See Information Sheet No. 5J for details.

SUMMARY

This chapter has described electronic circuit assembly. The following important points should have been learned:

- That electricians must be able to recognize and identify the different types of diagram used in electronics;

- How we use a *positional reference system* in electronic circuits to identify the location of a component;
- That BS 3939 symbols are used for the recognition of different types of electronic components on circuit diagrams;
- The basic principles of soldering and desoldering electronic components;
- The basic principles of wire wrapping and the advantages of this system of connection;
- The different types of circuit board used in electronic circuit assembly.

Test 5

Choose which one of the four answers is the correct one.

(1) The cross-section of terminal pins used for wire-wrapping is:

(a) triangular,
(b) pear shaped,
(c) square shaped,
(d) circular.

(2) When soldering, a crocodile clip is often used as a:

(a) rest for the soldering iron,
(b) heat shunt,
(c) component clamp,
(d) mains supply connector.

(3) The flux used in soldering helps to:

(a) avoid overheating of the component,
(b) keep the solder molten longer,
(c) clean the soldering iron,
(d) avoid oxidization of the joint.

(4) A copper desoldering braid is often used:

(a) as a heat shunt,
(b) as a component lead,
(c) to reduce the temperature when desoldering,
(d) to remove the solder from the joint.

(5) In electronics a positional reference system is used to:

(a) identify the location of components on a circuit board,
(b) find the manufacturer's reference number for a component,
(c) ensure that an IC is inserted the correct way round,
(d) help calibrate an oscilloscope to read peak voltage levels.

(6) Strip board consists of rows of copper strips bonded to:

(a) solder resist coated melamine board,
(b) synthetic resin bonded paper board,
(c) epoxy resin coated rigid polystyrene,
(d) compressed impregnated mica sheet.

(7) The symbols used in British circuit diagrams are drawn to:

(a) BS 1362,
(b) BS 3036,
(c) BS 3939,
(d) BS 3535.

(8) Which of the following diagrams is not used in electronics?

(a) schematic diagram,
(b) block diagram,
(c) circuit diagram,
(d) wiring diagram.

(9) A suitable etchant solution for PCB production would be:

(a) ferric chloride,
(b) acetic acid,
(c) citric acid,
(d) magnesium chloride.

(10) Tools that strip and wire-wrap the cable in one have:

(a) auto-stripping bits,
(b) cut-wrap bits,
(c) cut-strip-wrap bits,
(d) auto-wrap bits.

Chapter 6
Electronic Circuits and Systems

6.1 Systems

In the world of engineering there are many different systems - mechanical, electrical and electronic and very often a combination of these. For example, we could have a mechanical system for lubricating machinery, an electrical system that cuts off cable into 100 m lengths or an electronic system that will autofocus a security camera mounted near the entrance to a building. Generally speaking, systems fall into two main categories:

(1) The open loop system;
(2) The closed loop system.

The *open loop system* is one in which there is no monitoring of the output of any kind. Once turned on and operating, this system will continue to function regardless of any changes in the operating conditions. Consider a pump which is to fill a tank full of water. In the open loop system it is turned on and left to run, and it will continue to run even when the tank is full. This is not very clever, but it is an example of an open loop system where no monitoring takes place; not many of these systems are around.

With the *closed loop system* monitoring of the output takes place and the input is modified or adjusted depending on the state of the output. Consider our pump filling the the water tank again, but this time let it have a set of probes which are able to detect the level of the water in the tank. We can arrange these to switch off the pump when the correct water level is reached. Monitoring of the output is taking place and the input is being modified, or in this case stopped all together. This is a good example of the closed loop system in operation. Of course the system could be made fully automatic by including a second set of probes that will start the pump when the tank becomes empty.

Block diagrams

A number of block diagrams will be used in this chapter to illustrate the relationship between parts of the system under discussion and the arrows shown on certain ones indicate the general path of the signals/current flow between those parts. Block diagrams of both the open loop system and the closed loop system are shown on Information Sheet No. 6A.

Information sheet No.6.A.

Open and closed loop system

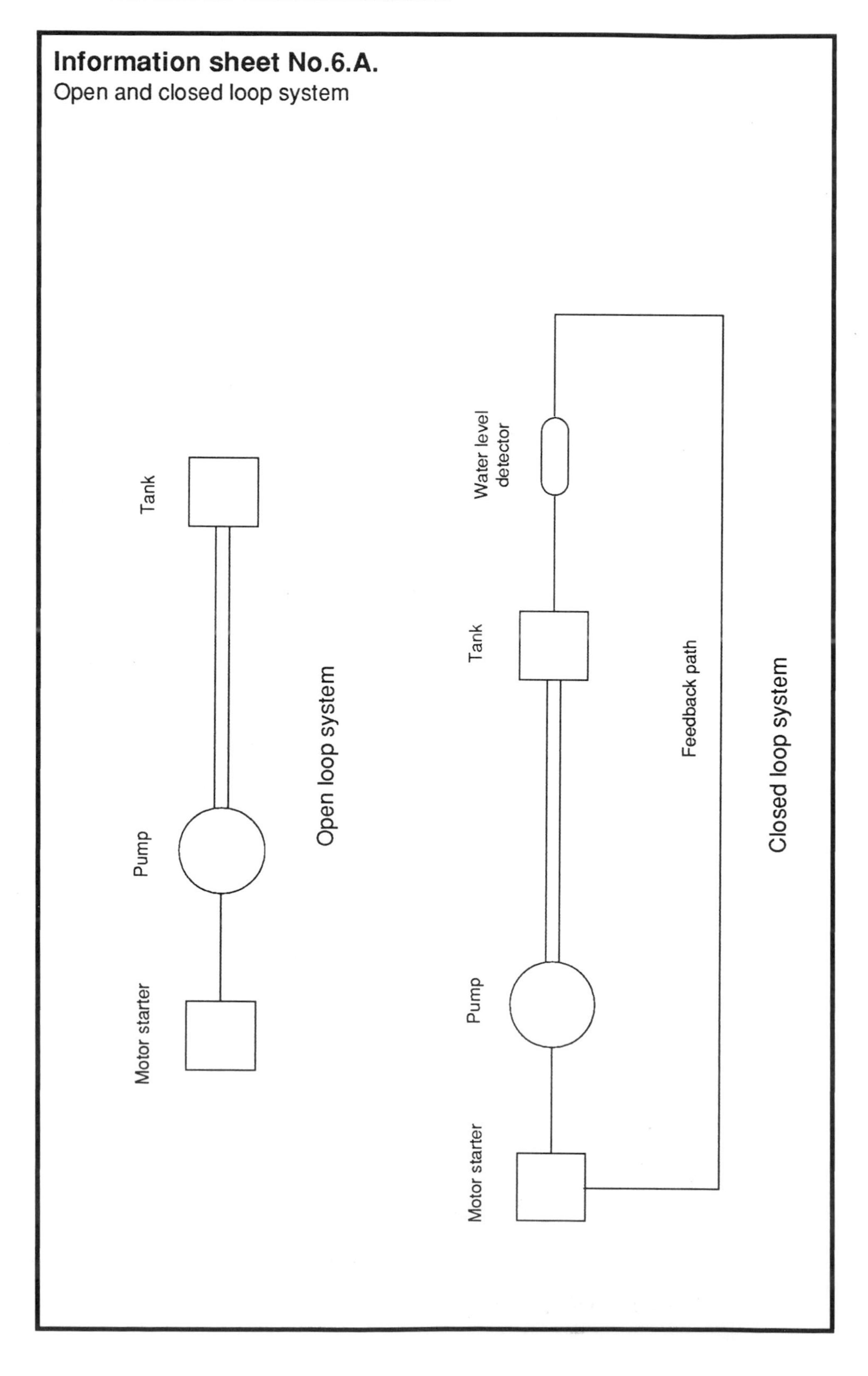

Feedback

Feedback in any system is where the output, or some portion of it, is fed back into the input of the system. The system does not have to be electronic, of course – it might well be human. Consider the driver of a car; the accelerator provides the input from the engine to the system, the speedometer provides the monitoring of the output from the engine to the car in terms of miles per hour, and the driver monitors the road conditions and provides the feedback link. If the car is going too fast the accelerator is released by the driver and the engine gives less power; if the car is going too slow the accelerator is depressed by the driver and the engine gives more power.

In an electronic system the circuits usually deal with signals and so the feedback will be in the form of a signal. Two types of feedback are possible:

(1) *Negative feedback* (sometimes called degenerative feedback); this occurs when the feedback signal opposes the input conditions.
(2) *Positive feedback* (sometimes called regenerative feedback); this occurs when the feedback signal enhances the input conditions.

Both these conditions are shown in Fig. 6.1.

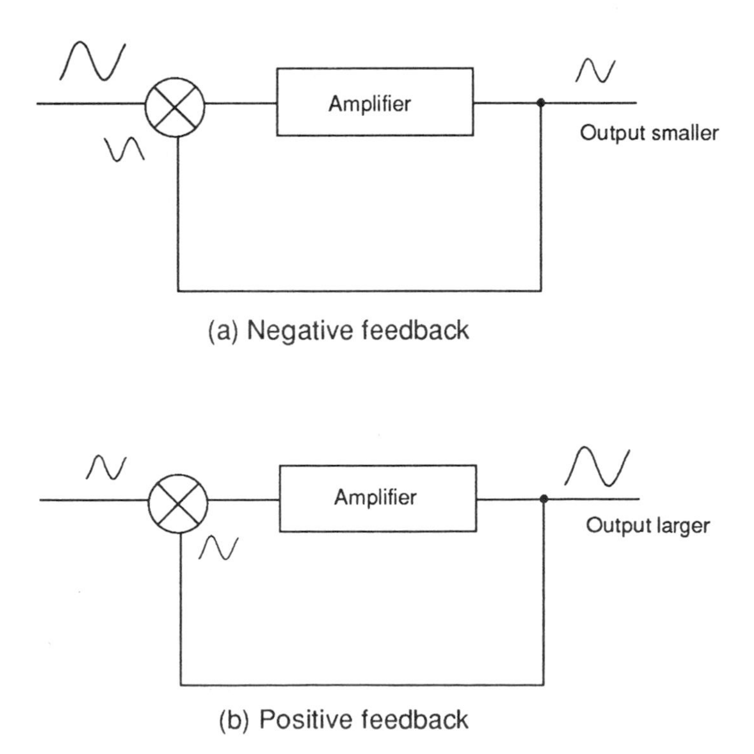

Fig. 6.1 Example of negative and positive feedback.

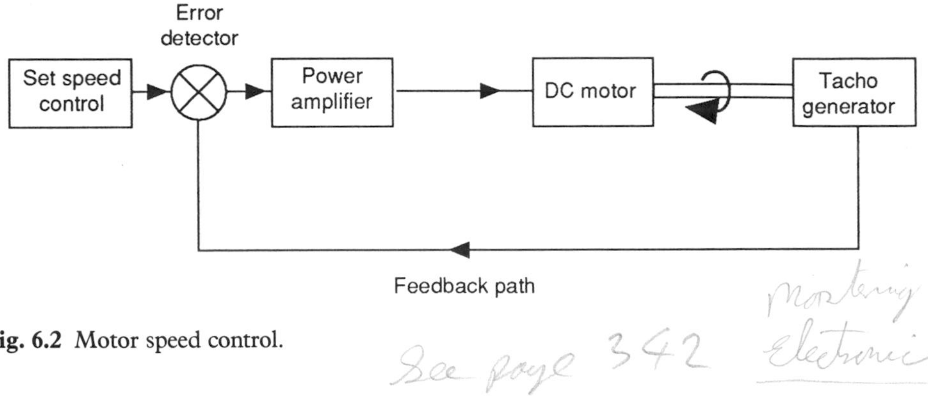

Fig. 6.2 Motor speed control.

6.2 Motor speed control

This illustrates how feedback may be used to control an electromechanical system. Consider the drawing in Fig. 6.2, which shows a typical block diagram of a speed control system for a d.c. electric motor.

The system consists of five distinct sections:

(1) *The set speed control*. This allows the operator to set the desired running speed for the motor, usually by the turn of a potentiometer.
(2) *The error detector*. This senses any differences in signal between the output from the tachogenerator and the set speed control.
(3) *The power amplifier*. This amplifies the signal from the detector, and feeds the signal to the d.c. motor.
(4) *The d.c. motor*. The speed of rotation depends upon the output from the amplifier. The motor also drives the tachogenerator.
(5) *The tachogenerator*. This produces an output signal directly proportional to the motor speed.

The tachogenerator is connected mechanically to the motor shaft, and the link from the tachogenerator to the the detector is the negative feedback path.

Imagine that the system is not energized, that the d.c. motor is not turning, and that there is therefore no output from the tachogenerator. The set speed control is at minimum, which means that no signal will be fed into the error detector section.

The set speed control is now adjusted to give a desired speed for the d.c. motor to run at – there will now be a difference in signal in the detector. This is amplified by the power amplifier, which in turn drives the motor. The tachogenerator will now give an output depending on the speed of the d.c. motor. This is fed back to the detector, the difference in signal is again amplified, and the d.c. motor continues to increase in speed. Naturally, this cycle repeats itself until the output from the tachogenerator and the signal from the set speed control are identical. The error detector senses that the signals are the same and the power amplifier is now at the correct level to keep the motor rotating at the desired speed.

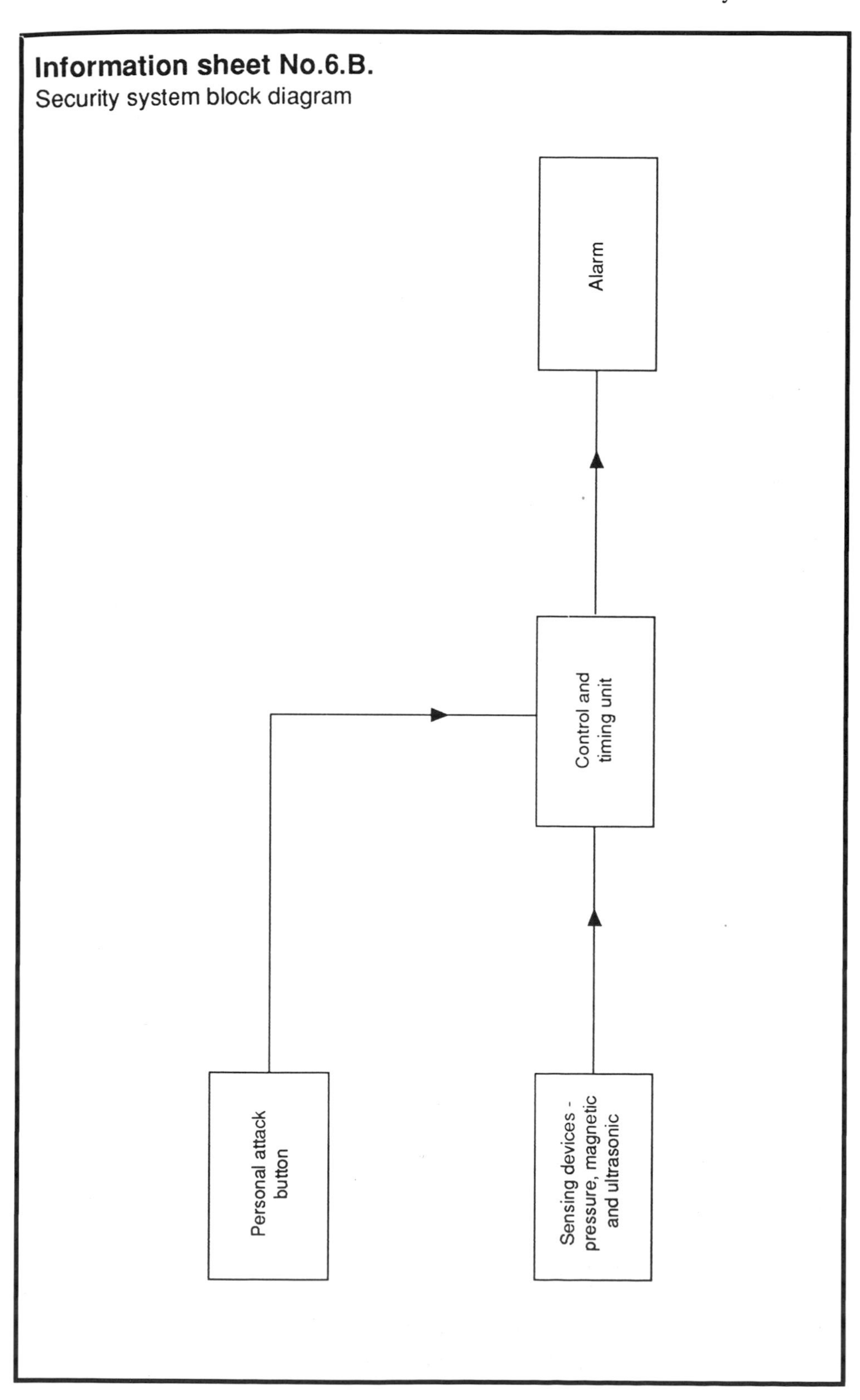
Information sheet No.6.B.
Security system block diagram
Alarm
Control and timing unit
Personal attack button
Sensing devices - pressure, magnetic and ultrasonic

If the set speed control is further adjusted, either up or down, then again the detector will sense this difference, the power amplifier either increasing or decreasing its output as required.

6.3 Security protection systems

The block diagram shown on Information Sheet No. 6B is for a typical security system, and incorporates the following units:

(1) *Personal attack button.* This is a manually operated device for the protection of the individual. It is much used in places such as banks, social security offices, bookmakers, etc. Once activated, it sends a signal to the controller, and the alarm is sounded.
(2) *Sensing devices.* These are automatic in their operation and can be of the magnetic, ultrasonic or pressure types. Once activated, the device sends a signal to the controller and the alarm is sounded.
(3) *The control unit.* This receives the signals from the other devices and sounds the alarm.
(4) *The alarm sounder.* This can be a bell, buzzer, siren or klaxon horn; it is set off by the control unit once it has received a signal from one of the sensing devices. More sophisticated systems will not sound an alarm but will send a signal direct to the police or a security organization.

Security system circuit diagram

The circuit diagram on Information Sheet No. 6C incorporates a number of electronic devices that we have discussed in earlier chapters; these should be quite easily understood by the student.

Operating the personal alarm button, activating any of the sensing devices, or cutting the wires to any of these devices in the external circuit will make the alarm sound.

When the circuit is broken, a current flows via the 22 000 Ω resistor, through the cathode-gate junction of the thyristor. This switches on the thyristor, allowing current to flow through the alarm sounder via the anode–cathode circuit. To stop the alarm, the circuit should be made continuous again and the mute/reset switch on the alarm unit switched off momentarily.

Sounders such as electric bells and buzzers have make and break contacts built in them which automatically switch the current on and off. Activating one of the sensors or operating the personal alarm button would break the external circuit. This would start the alarm, but the current to the thyristor would be switched off by the make and break contact in the sounder as it opened and closed, and so the the alarm would cease. In order to ensure that the alarm continues to sound, the

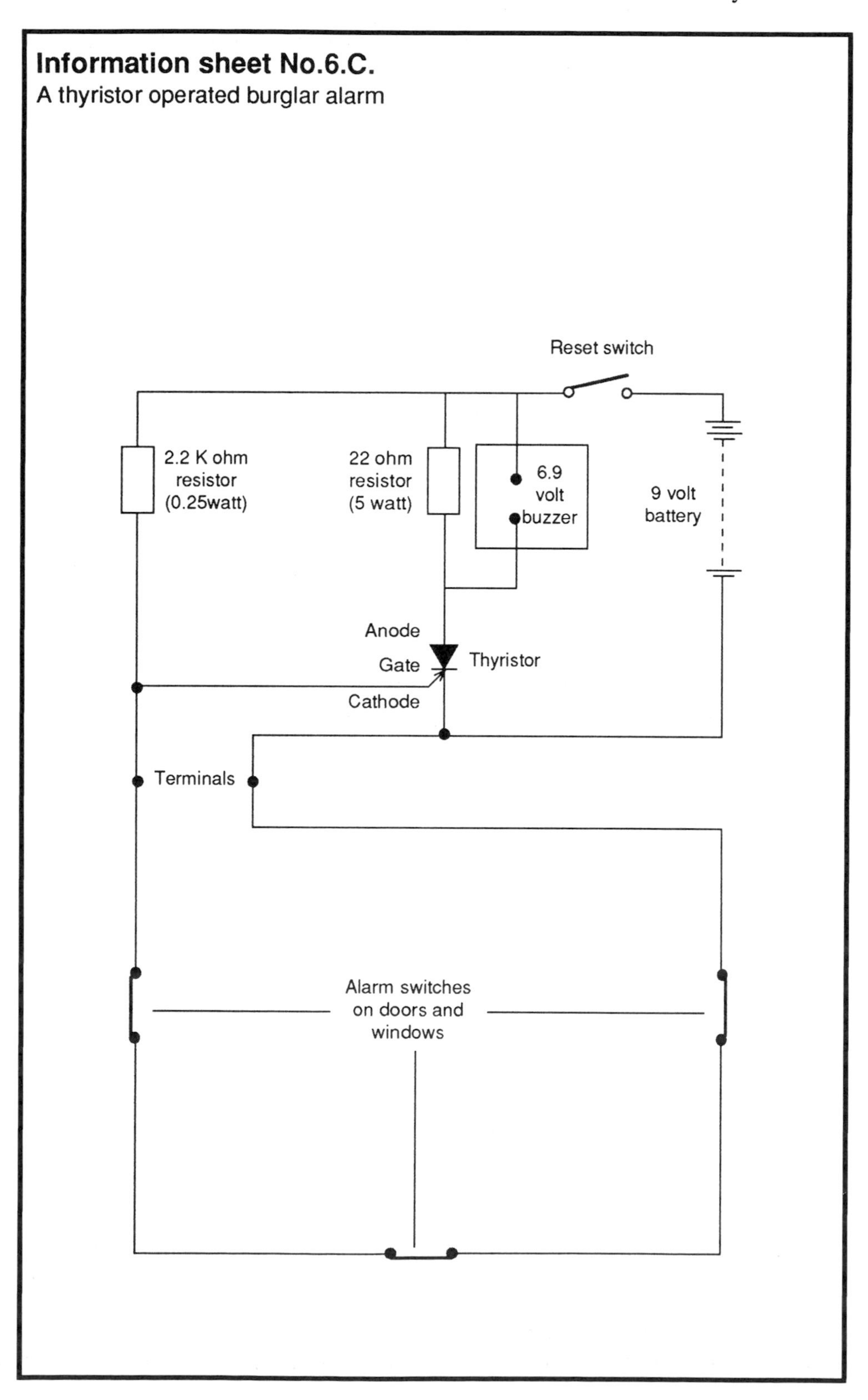
Information sheet No.6.C.
A thyristor operated burglar alarm
Reset switch
2.2 K ohm
resistor
(0.25watt)
22 ohm
resistor
(5 watt)
6.9
volt
buzzer
9 volt
battery
Anode
Gate
Thyristor
Cathode
Terminals
Alarm switches
on doors and
windows

circuit which includes the anode and cathode of the thyristor must be made continuous. This is ensured by the 22 000 Ω resistor connected across the sounder. Even when the current to the thyristor is interrupted by the contacts in the sounder, current still flows via the resistor. The thyristor used is a CRS/025 AF; with any other type the value of the resistor connected to the gate may have to be reduced.

Types of detector and their application

Infra-red Beam This consists of a transmitter unit emitting a pulse modulated beam and a receiver unit with a parabolic receiving area giving them an acceptance angle of up to 20°. The pulse modulated beam is activated into the alarm condition when the invisible infra-red ray is interrupted. The units can be operated at any distance up to 15 m apart and comply with BS 4737: Part 3.

Passive infra-red These units consist of a sophisticated double mirror and twin element pyroelectric sensor which converts changes in infra-red radiation caused by people or other sources of heat into electronic signals for alarm processing. Some units can be programmed for either one- or two-event operation, i.e. initiate an alarm on detection or go into a standby mode and raise the alarm if a second event occurs within 33 seconds.

Combined microwave/passive infra-red Detectors of this type combine a microwave motion detector with a passive infra-red detector sensing heat radiating bodies. The infra-red radiation given off by the heat source is focused onto the infra-red radiation detector by a Fresnel lens and in some models a mirror is incorporated to detect objects immediately under the detector. The two units need to sense an intruder simultaneously before an alarm state is triggered, so false alarms are very unlikely.

Ultrasonic motion detector This detector works on a patented balanced signal processing system which recognizes the difference between random motion and intruder motion, thus providing a high degree of immunity from false alarms. The detector generates a three-dimensional oval volume detection area which may be adjusted by a range control up to a distance of 9 m. A number of detectors can be fitted in the same room without needing synchronization. The unit is suitable for use with BS 4737 alarm control systems.

Proximity switches There are numerous types of proximity switch both for fitting flush into the woodwork of doors or windows or mounting on the surface. Most models consist of two units, one a reed switch contained in a moulded plastic case which is mounted in the door jamb and the other a permanent magnet mounted in the door exactly opposite the reed switch. When the door is closed the permanent magnet keeps the switch open or closed (depending on whether a

Information sheet No.6.D.

Washer block diagram

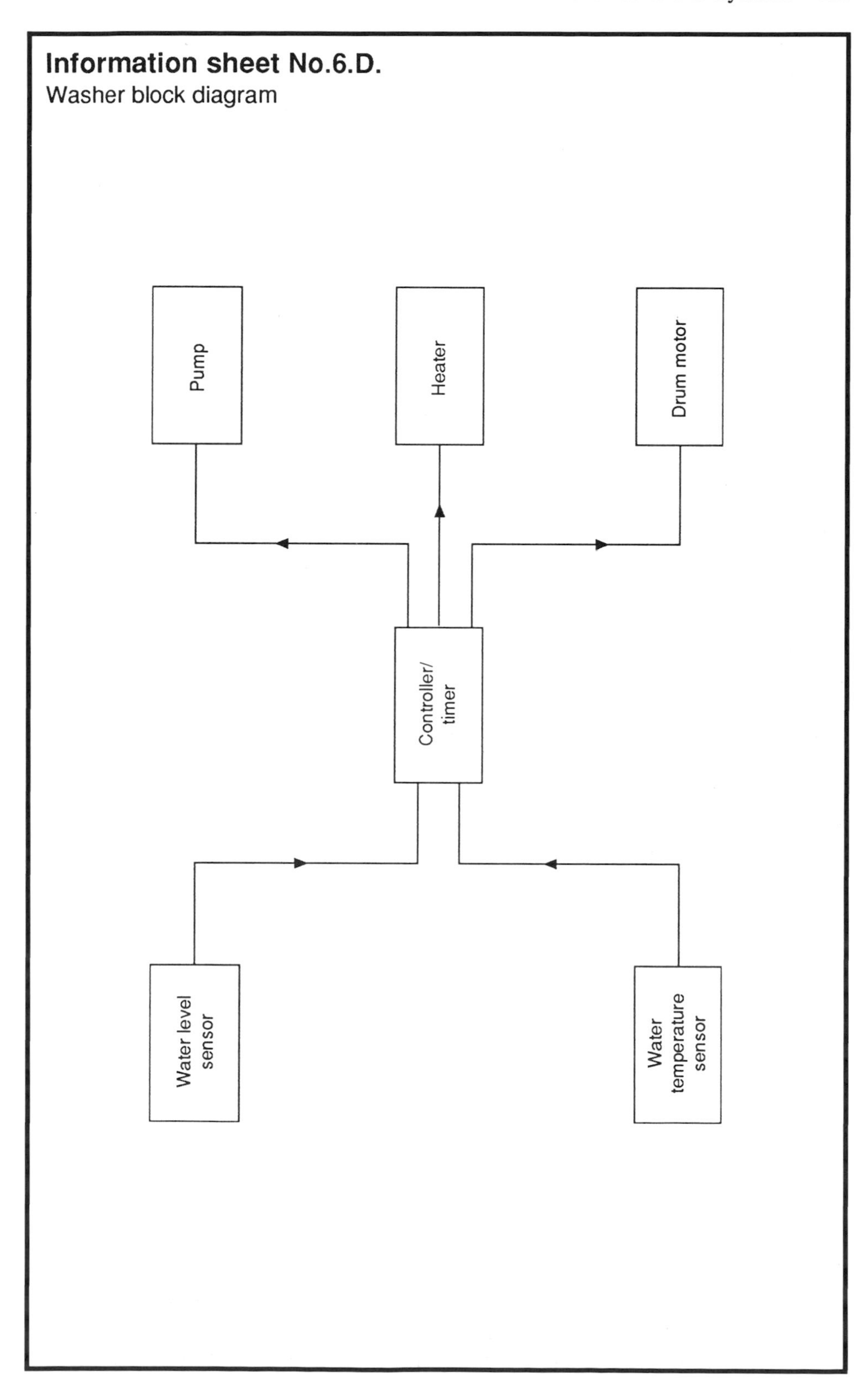

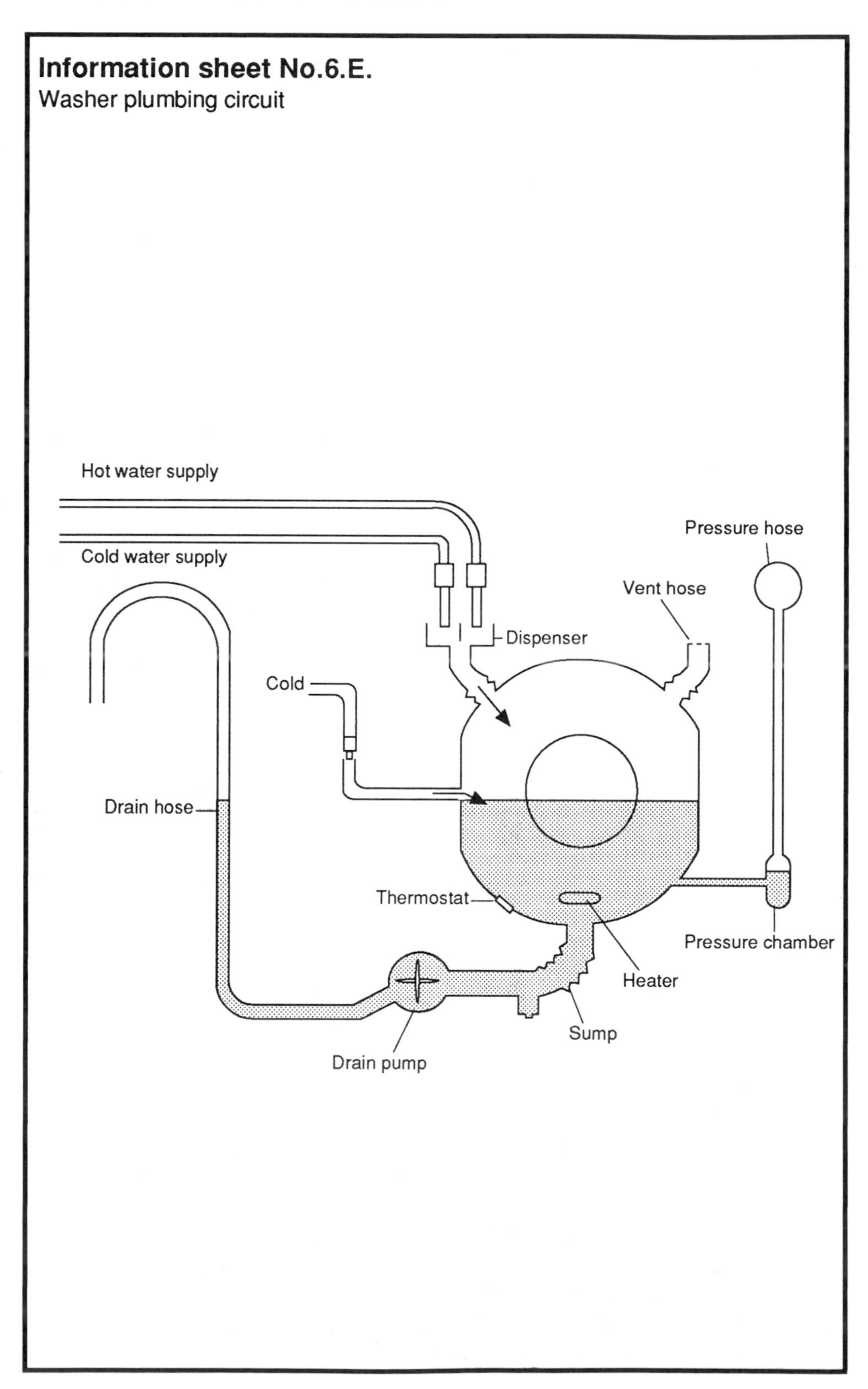
Information sheet No.6.E.
Washer plumbing circuit
Hot water supply
Cold water supply
Pressure hose
Vent hose
Dispenser
Cold
Drain hose
Thermostat
Pressure chamber
Heater
Sump
Drain pump

closed or open circuit is to be used). On opening the door the contacts are no longer under the influence of the magnet and the alarm is sounded.

Pressure pad switches These usually take the form of a thin sealed heavy plastic pad suitable for placing under carpets near doors or windows or under stair carpets. The pad, which is often referred to as a *resilistor*, contains carbon granules. The weight of a person stepping onto the pad compresses the granules, reducing their resistance and activating the alarm circuit.

6.4 Washing machine control circuits

The block diagram shown on Information Sheet No. 6D is for a typical washing machine control system, and incorporates the following units:

(1) *Water level sensor*. This signals the controller when a selected water level has been reached. In a good number of washing machines these are operated by a pressure switch, as shown in the plumbing diagram on Information Sheet No. 6E. However, a water level detector can be a simple float switch, or a device in which a variable resistor is operated, so changing the resistance and thus indicating the level of the water. The water level indicator is incorporated into the washing machine circuit in such a way that the machine will not operate unless the water level (and this includes the absence of water altogether) for a particular operation is correct.
(2) *Temperature sensing devices*. These are automatic in their operation and will indicate to the controller when a selected temperature has been reached. In older washing machines this would be a type of thermostat operating on the bi-metal strip principle. More modern machines will have full electronic circuitry control incorporating a thermistor. The advantage of the thermistor is its rapid response to temperature change.
(3) *The programme timer/control module*. This receives the signals from the sensing devices and operates the washing machine drum motor, the heater and the water pumps in accordance with the timer/selection switch. Its operation is explained below.
(4) *The water pump*. This pumps water in or out of the tub as required by the controller.
(5) *The water heater*. If the temperature sensing device detects that the water is not up to the required temperature the water heater will be brought into operation by the controller and the wash will not proceed until this temperature has been reached. If the water entering the washing machine is already hot this will be detected by the temperature sensing device and the water heater will not be brought into play.
(6) *The drum motor*. The speed of the drum motor is governed by the control module . A full description of how this operates is given later.

Information sheet No.6.F.
Circuit one

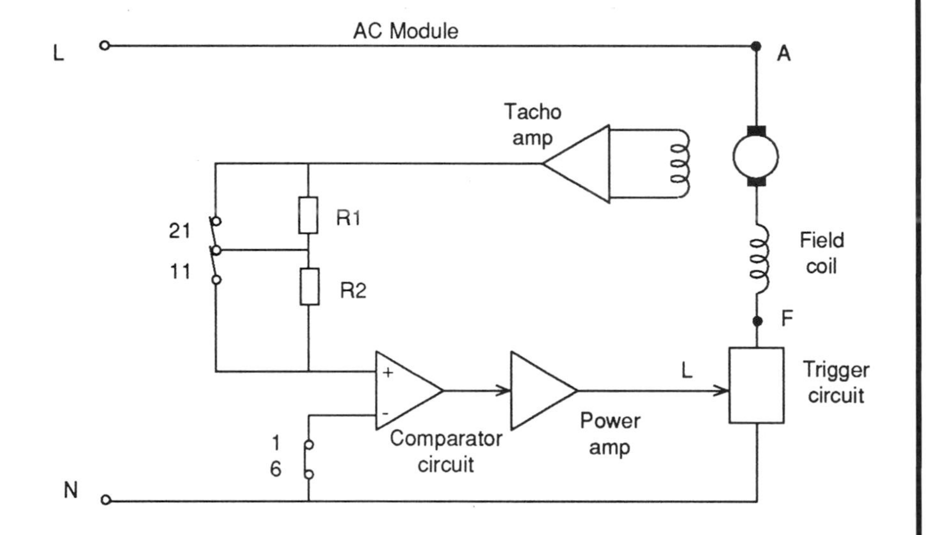

Information sheet No.6.G.

Circuit two basic module diagram

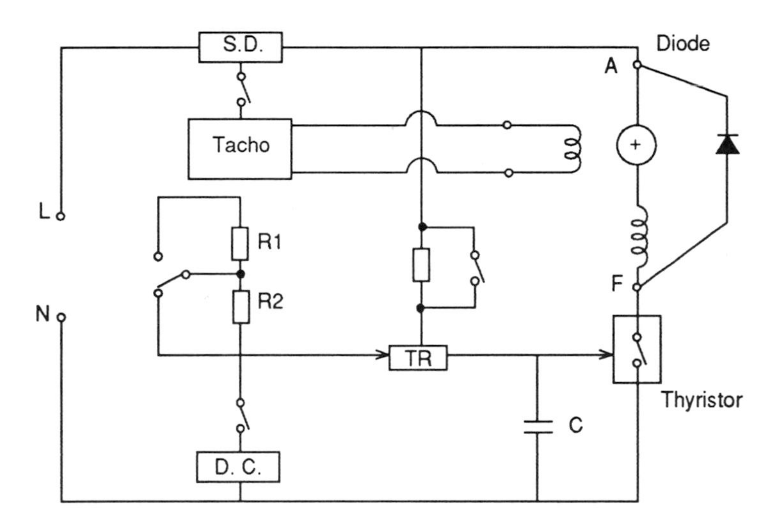

Washing machine circuit diagrams

The circuit diagrams on Information Sheets Nos. 6F and 6G incorporate a number of electronic devices that we have discussed in earlier chapters, and the student should be able to understand them with little difficulty.

The programme timer

This is basically a motor-driven switch with a large number of different positions. It also incorporates an axial on/off switch which can be operated manually. The switch drives a cam which operates switches bringing in the different stages of the wash. To move from one position to the next, i.e. impulse, the limits on the present position must be satisfied. The relevant control component (pressure switch, thermostat, time, etc.) must complete a circuit. This must be either direct to the timer, or to a series of minute timer contacts, if several conditions have to be satisfied before the cam is turned into the next position.

Control module

There are two main methods of controlling the speed of the drum motor:

(1) By the use of a tachogenerator and electronic circuitry incorporating resistors as the speed-controlling components;
(2) By the use, again, of a tachogenerator, but using electronic circuitry incorporating a thyristor as a *wave chopping* device.

Method 1

The circuit diagram for this first method is shown on Information Sheet No. 6F. This method makes use of a small tachogenerator fixed to the rear end of the motor shaft; this provides feedback voltage proportional to the speed of the motor, i.e. the higher the speed the higher the generated feedback voltage.

This voltage is boosted by an amplifier circuit within the module and then fed through two resistors. By bypassing one, two or neither of the resistors we can obtain the three speeds: spin, distribute and tumble.

The voltage feedback from the generator is then passed into a comparator where it opposes a negative supply obtained from the main gate supply. Some of the main gate supply is then cancelled by the tacho-voltage and the remainder is used to trigger the circuits that control the current passing through the motor. Therefore, the higher the resultant voltage from the comparator the faster the motor speed, and vice versa.

The control of the spin, distribute and tumble speeds operates as follows:

- With both the resistors in circuit the tacho-voltage is reduced to a minimum, therefore giving no opposition to the main gate supply, and the resultant voltage is maximum and gives spin speed.

- By bypassing just one resistor the resultant voltage is less and the speed is reduced, i.e. distribute speed.
- With both resistors bypassed the resultant voltage is at its minimum and the speed is low, i.e. tumble speed.

On the two lower speeds, control is constant so if the speed rises the tacho-voltage rises and opposes the main gate supply, thus reducing the resultant voltage and so dropping the speed again. The opposite is also true: if the speed falls the tacho-voltage falls and and gives less opposition to the main gate supply – thus the resultant voltage is greater and the motor speeds up. This is a continuous process; see the circuit diagram on Information Sheet No. 6F for further details.

Method 2

The circuit diagram for this second method is shown on Information Sheet No. 6G. As before, the module is required to control the motor to three speeds: spin, distribute and tumble.

Because of the diode action of the thyristor, the supply to the motor is d.c. (more correctly unidirectional a.c.) as shown in Fig. 6.3. No current is passed to the motor between points *a* and *b*; therefore, the earliest point that the thyristor needs to be switched on is *b*. Then the whole half cycle of power could be passed on to the motor.

To obtain maximum speed for spin, we require all the power that is available. We would like this to commence at point *b* but, because of the delay in switching time (which will be explained later), the earliest starting point is at *z*. If the switching time is delayed to point *y*, power will only be supplied from *y* to *c*. This represents only a small part of the total and therefore the motor will run at reduced speed, i.e. distribute speed. By delaying the switching time even further to point *x* the power will only be available from *x* to *c* and the speed will be at its lowest, i.e. tumble speed.

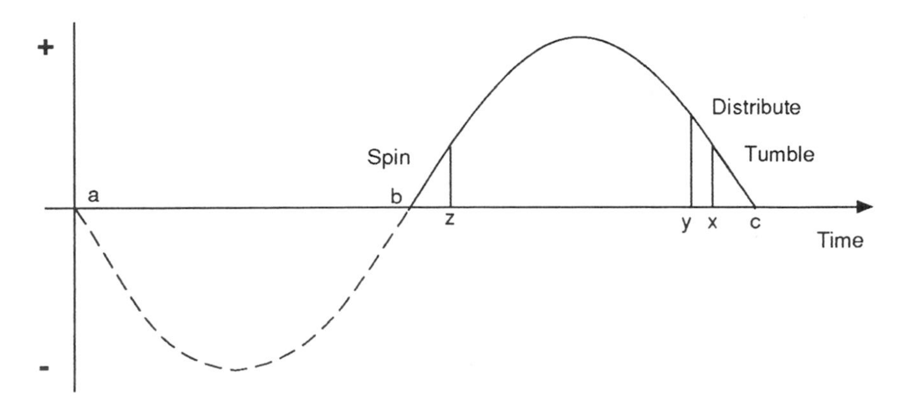

Fig. 6.3 Wave chopping.

Information sheet No.6.H.

DC power supply

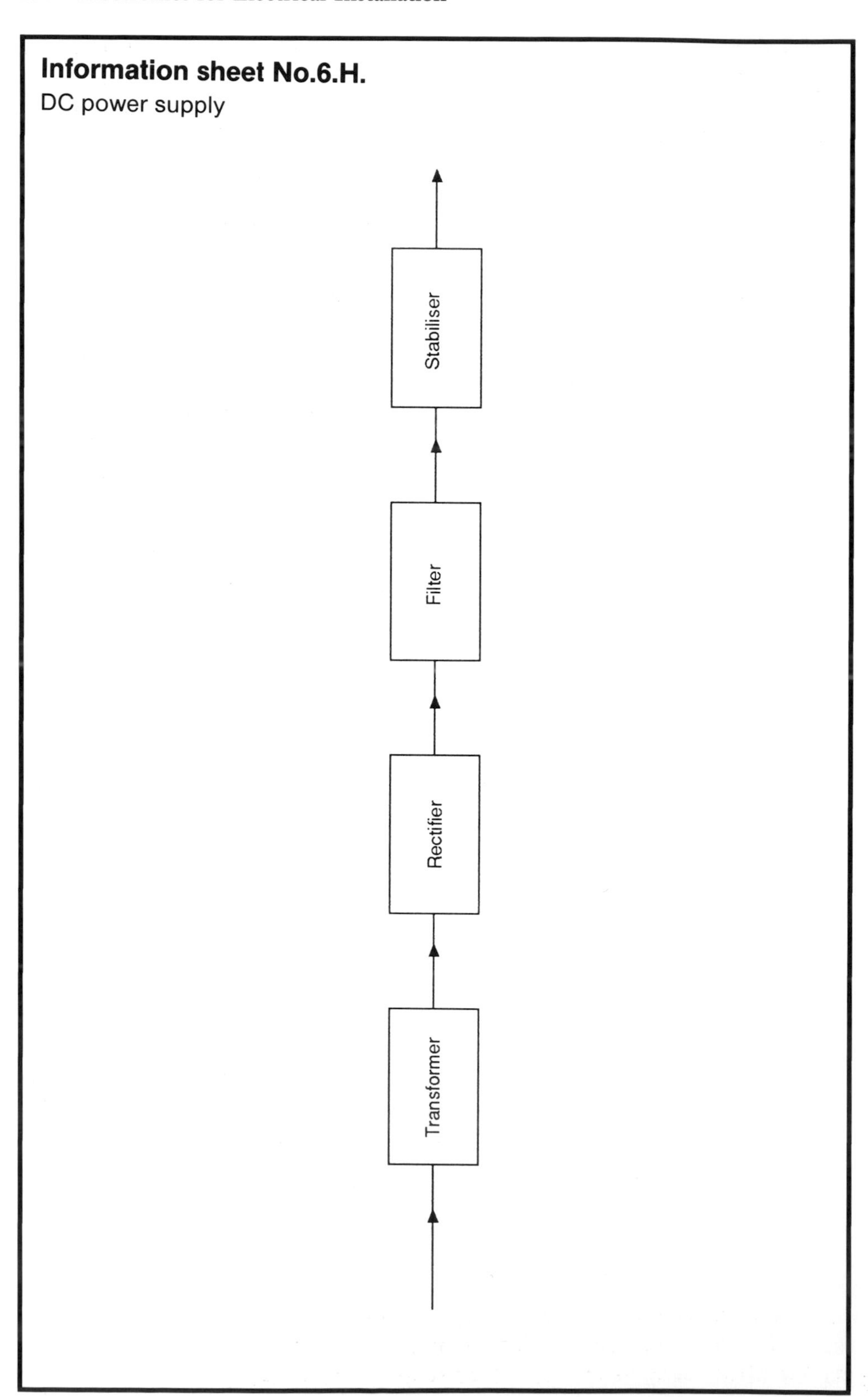

The above describes the power circuit of the module, but we also require provision for a switching voltage and time delay. The switching voltage is obtained from the capacitor. On switching on initially, the capacitor is uncharged. After a period of time when the capacitor becomes fully charged, the thyristor is triggered and the motor runs. This delay in charging explains why we cannot get all the power from the half cycle by starting at *b*. From this it can be seen that by varying the charge rate we can control the power fed to the motor and hence its speed.

To obtain a controlled low speed a tachogenerator is utilized in much the same way as in Method 1; here it produces a voltage in opposition to the d.c. supply and because it is connected to the motor shaft this varies in direct relation to the speed of the motor. If a proportion of the generated feedback voltage is connected to the d.c. supply, this will have the effect of reducing the charging rate of the capacitor. This delays the switching on of the thyristor and thereby gives a lower speed. The opposite is also true; if the tacho-voltage falls due to a drop in the motor speed the charging rate is no longer reduced and the motor speeds up. In this way the motor is kept at a constant speed.

Because the fluctuations of the tacho-generator are so small in magnitude we amplify them with a transistor circuit. In this way, the small variations are used to switch a larger voltage for use in charging the capacitor.

6.5 D.C. power supplies

The block diagram shown on Information Sheet No. 6H is for a typical d.c. power supply system, and incorporates the following units:

- Transformer;
- Rectifier;
- Filter;
- Stabilizer.

In electronics, the phrase *power supply* is loosely used to describe a circuit that will convert the 240 V alternating waveform of the mains supply into an extra low voltage (say 12 V) d.c. supply to power an electronic circuit.

Rectification means changing from a.c. to d.c., and is an old expression used when d.c. mains supplies were common.

There are several circuit arrangements to achieve rectification, and broadly speaking these are:

(1) Half wave rectification;
(2) Full wave rectification, centre tapped transformer method;
(3) Full wave rectification, bridge method;
(4) Switched mode; this is beyond the scope of this book.

Information sheet No.6.I.
Half wave rectification

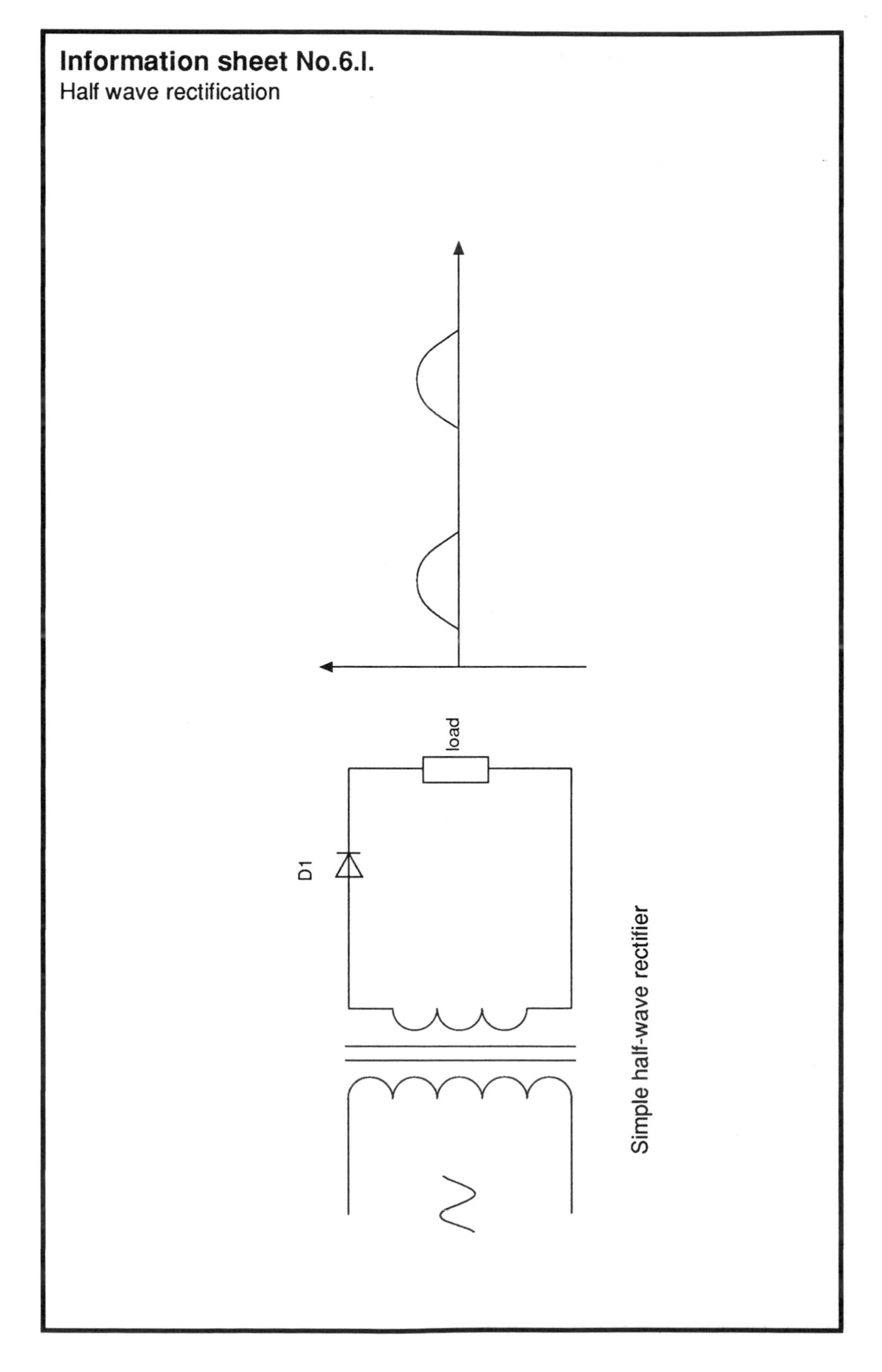

Simple half-wave rectifier

Information sheet No.6.J.

Full wave rectification using a centre-tapped transformer

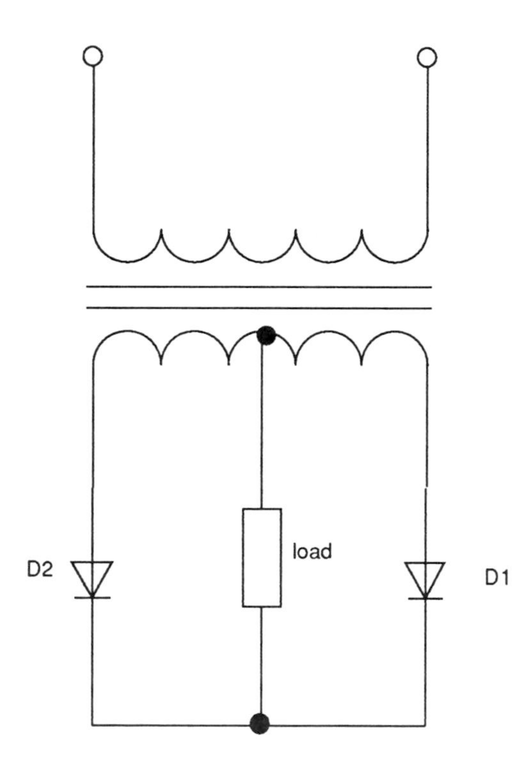

Half wave rectification

The half wave rectifier is very basic; it simply consists of a transformer, and a single diode. This circuit and the voltage output waveform is shown on Information Sheet No. 6I.

It is called half wave because, as the diode only conducts in one direction, only the positive half of the waveform is present. There is a space where the negative half wave should be. This circuit is very crude and is suitable for only the most basic of circuits such as battery chargers.

Full wave rectification

An improvement on half wave rectification can be obtained by the use of two diodes and a centre tapped transformer. This is suitable for more complex equipment and the circuit diagram for this is shown on Information Sheet No. 6J.

Diode D1 conducts for the positive half cycle of the wave, diode D2 for the negative half cycle of the wave. Notice that the load is connected to the centre-tap of the transformer. The big disadvantage of this type of rectification system is that the output voltage of the transformer has to be double that of the load.

Full wave bridge rectifier

This circuit is superior to the others mentioned, and makes use of the full waveform. The basic circuit is shown on Information Sheet No. 6K.

The circuit consists of the following components:

(1) A step-down transformer to reduce the voltage from 240 V to the required voltage for our circuit;
(2) A pair of fuses (d.c. is fused in both poles);
(3) A bridge rectifier comprising four diodes connected in a diamond shape and sealed in a single package;
(4) A load, represented by the resistor.

If the bridge was to be made up from discrete components it would be essential to ensure that the diodes were connected in the correct configuration or the circuit would not work.

From the secondary side of the transformer, connections are made to two opposite points on the diamond shape, as shown in the circuit diagram on Information Sheet No. 6K. A second set of connections is then taken from the remaining two points and connected to the load.

Let us consider what happens when the output from the secondary side of the transformer is a positive waveform. The path taken by the current will be through

Information sheet No.6.K.

Full wave rectification using a bridge rectifier

Information sheet No.6.L.

Current paths through bridge rectifier for each half cycle

Circuit during positive half wave

Circuit during negative half cycle

Output voltage waveform for full wave rectifier

D1, and, as D2 is not conducting, it must go through the load and back to the transformer via D4.

Now let us consider what happens when the output from the secondary side of the transformer changes to a negative waveform. Here the current path will be through D2, and, as D1 is not conducting, it must go through the load and back to the transformer via D3.

It should be noted that, despite the change from the positive half cycle to the negative half cycle, the bridge still manages to send the current to the same leg of the load each time, so making it a unidirectional supply. In this form the output is referred to as *raw d.c.*. This raw d.c. can be made smoother by incorporating a smoothing circuit and further details of this can be found in Chapter 8. The output waveform along with the current path for both half cycles is shown on Information Sheet No. 6L.

Filter circuits

So far we have seen that the alternating high voltage can be rectified to provide a low voltage smooth d.c. supply for the load. There are many applications where this is not enough. In electrically noisy environments it is likely that mains-borne high voltage, high frequency spikes (sometimes called transients) may be present in the supply. To prevent these spikes being transferred to our power supply a low pass filter is incorporated into the circuit after the bridge rectifier. This has the effect of filtering out the unwanted transients. In highly complex power supplies it might be necessary to fit a filter on the high voltage mains side of the transformer.

Further information on d.c. supplies can be found in Chapter 8 of this book, and details of the European Directive on Magnetic Interference, which lays down the law on the magnetic compatibility of equipment, can be found in *Electrical Installation Practice, Book Three.*

Stabilizing circuits

It is highly desirable that, should the load on the power supply be increased (thus drawing more current) or, equally, the voltage from the rectifier rise for some reason, the output voltage remains stable.

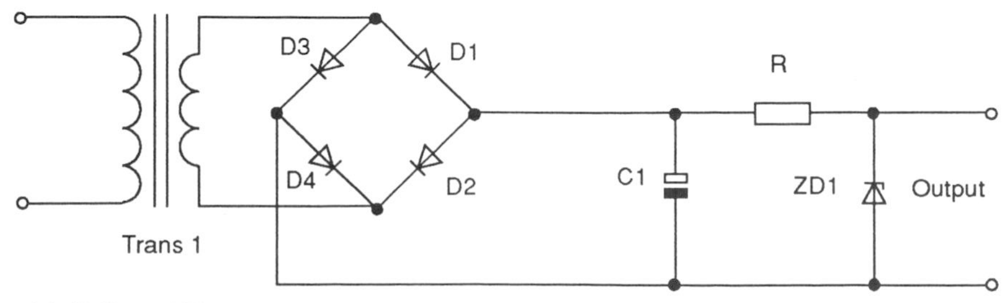

Fig. 6.4 Fully stabilized power supply.

Information sheet No.6.M.

Space heating block diagram

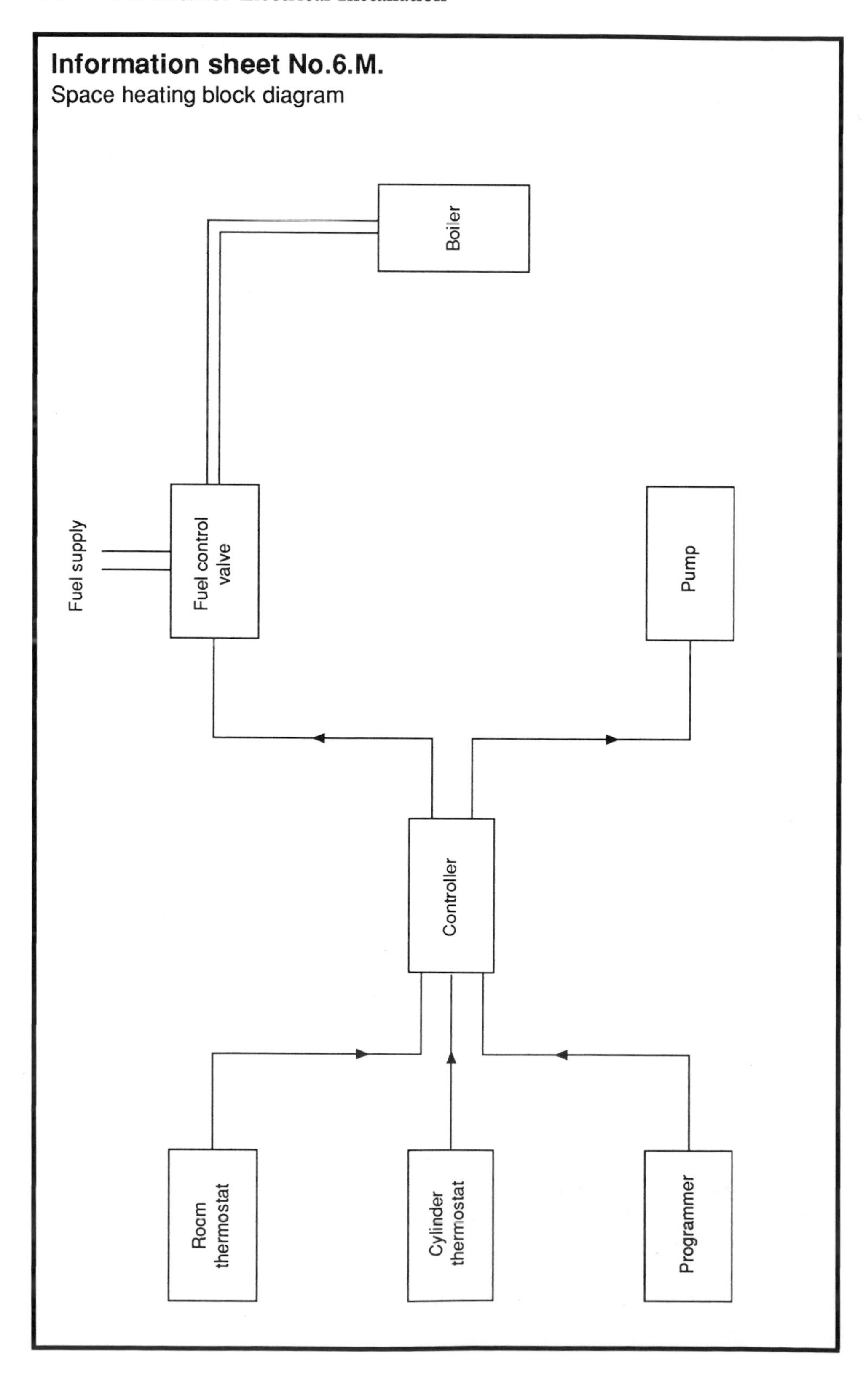

To provide this facility the power supply is further refined by the inclusion of a stabilizing circuit. The circuit diagram for this is shown in Fig. 6.4; it incorporates a number of electronic devices that we have discussed in earlier chapters and should be understood with little difficulty by the student.

The zener diode acts as a voltage reference, and once breakdown occurs the voltage across the diode remains almost constant, regardless of the current through the device. Thus, if either the load current or the supply voltage should vary in any way, then the zener diode current will change and so will the voltage drop across the series resistor, keeping the output voltage at a constant value. Further information on stabilization can be found in Chapter Eight.

6.6 Space heating systems

The block diagram shown on Information Sheet No. 6M is for a typical space heating system, and incorporates the following units:

(1) Room thermostat;
(2) Cylinder thermostat;
(3) Programmer;
(4) Control unit;
(5) Fuel control valve;
(6) Pump;
(7) Boiler.

The room thermostat This is really a switch that reacts to temperature change. It consists of a small box about the size of a one-way lighting switch which has a temperature selection dial on the front. The dial is marked off in degrees (centigrade in recent times). The thermostat uses the bimetalic strip principle of operation which is arranged to *make* a set of contacts when cold and *break* them when it gets hot. The temperature selection dial simply moves a cam arrangement, which makes it either easier for this to happen when more heating is required, or harder for it to happen if less heating is required.

The cylinder thermostat This work in exactly the same way as the room thermostat above; it is, however, usually of more robust construction. It will either be fastened in close contact to the outside of the copper cylinder by means of a stout strap, or inserted in a special pocket in the cylinder body. The thermostat is used to sense the temperature in the cylinder and in conjunction with a diverter or motorized valve may be used to control the domestic hot water temperature; also, where required, it is possible to secure the priority of the temperature of hot water needs in preference to space heating demands. Although temperature adjustment is provided, a cylinder thermostat is normally set to cut off at 60°C (140°F).

The programmer This automatically switches on the heating system at times suitable to the user's needs. Dependent on the complexity of the device it is possible to provide a basic set pattern of on/off operations using a simple time switch to a quite complex arrangement using a time switch in conjunction with multiple manually operated selection switches. Here, the user can select one of a number of programmed sequences of operation, i.e. space heating, domestic hot water or both.

The control unit This is really at the centre of things; it serves as a place to which all the various sensing devices, pumps, programmer, etc., can be wired back, terminated and interconnected. It will also contain such things as relays, electronic circuitry and protective devices. These units can sometimes form part of the programmer, or in the case of ducted hot air systems are actually built into the heater unit.

The fuel control valve This is an electromagnetically operated solenoid valve, piped into the fuel supply line. With the programmer in the *on* position and the thermostats calling for heat, the solenoid will operate and allow fuel to flow to the boiler's burner. The valves are interconnected to a flame-sensing device which might be a thermocouple device, as discussed in Chapter 1, or a photoelectric device which sees the flame. Either device will cut off the electrical supply to the solenoid valve on flame failure, so shutting down the fuel supply.

The pump For hydraulic (wet) systems it is now common practice to use a pump (often referred to as an accelerator) to propel the water round the system. This obviates the need for reliance on the gravity displacement effect. A pump is normally designed to operate at the 240 V, 50 Hz, mains supply and the single phase induction motor seems to be most favoured.

Systems may be designed with the pump operating continuously (essential on solid fuel systems) or with intermittent pump operation under the control of any or all of the following: a room thermostat, a time switch, a programmer or a manual selection switch.

The boiler Whether fired by solid fuel, oil or gas the boiler's purpose is to provide the heat for both the domestic hot water and the heating system. It will incorporate a flame failure device which will shut down the fuel supply to the burner if no flame is present. It may have a pilot light which burns whenever the system is in operation, or it might rely for ignition on a spark generator which operates only on demand.

Space heating diagrams

The wiring diagram (a) on Information Sheet No. 6N shows the external connections for a typical 16-way programmer. The programmer is designed to

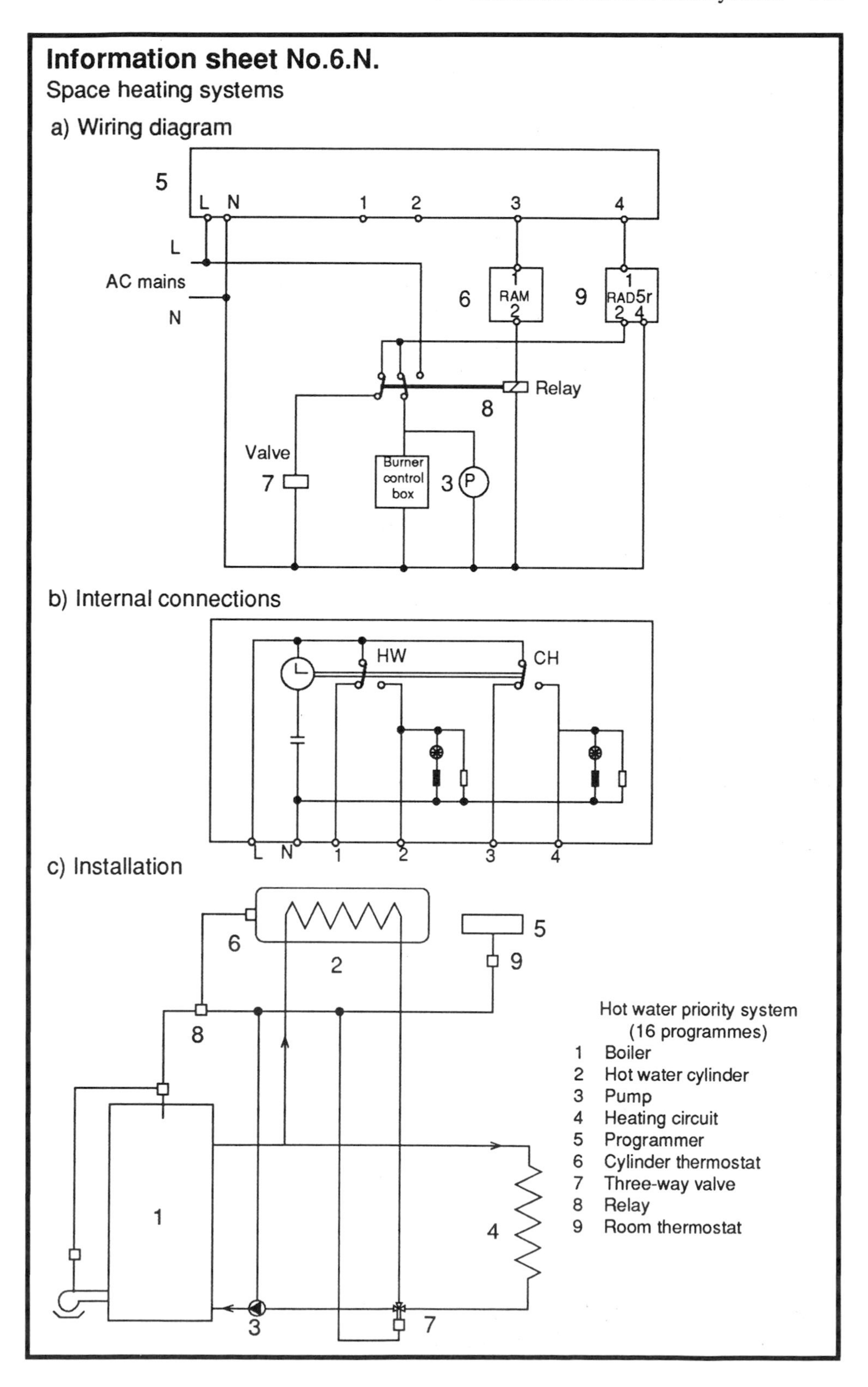
Information sheet No.6.N.
Space heating systems
a) Wiring diagram
5
L N 1 2 3 4
L
AC mains
N
6
RAM
9
RAD5r
Relay
8
Valve
7
Burner control box
3
P
b) Internal connections
HW
CH
L N 1 2 3 4
c) Installation
6
2
5
9
8
1
4
3
7
Hot water priority system (16 programmes)
1 Boiler
2 Hot water cylinder
3 Pump
4 Heating circuit
5 Programmer
6 Cylinder thermostat
7 Three-way valve
8 Relay
9 Room thermostat

Information sheet No.6.O.

Lamp dimmer circuit

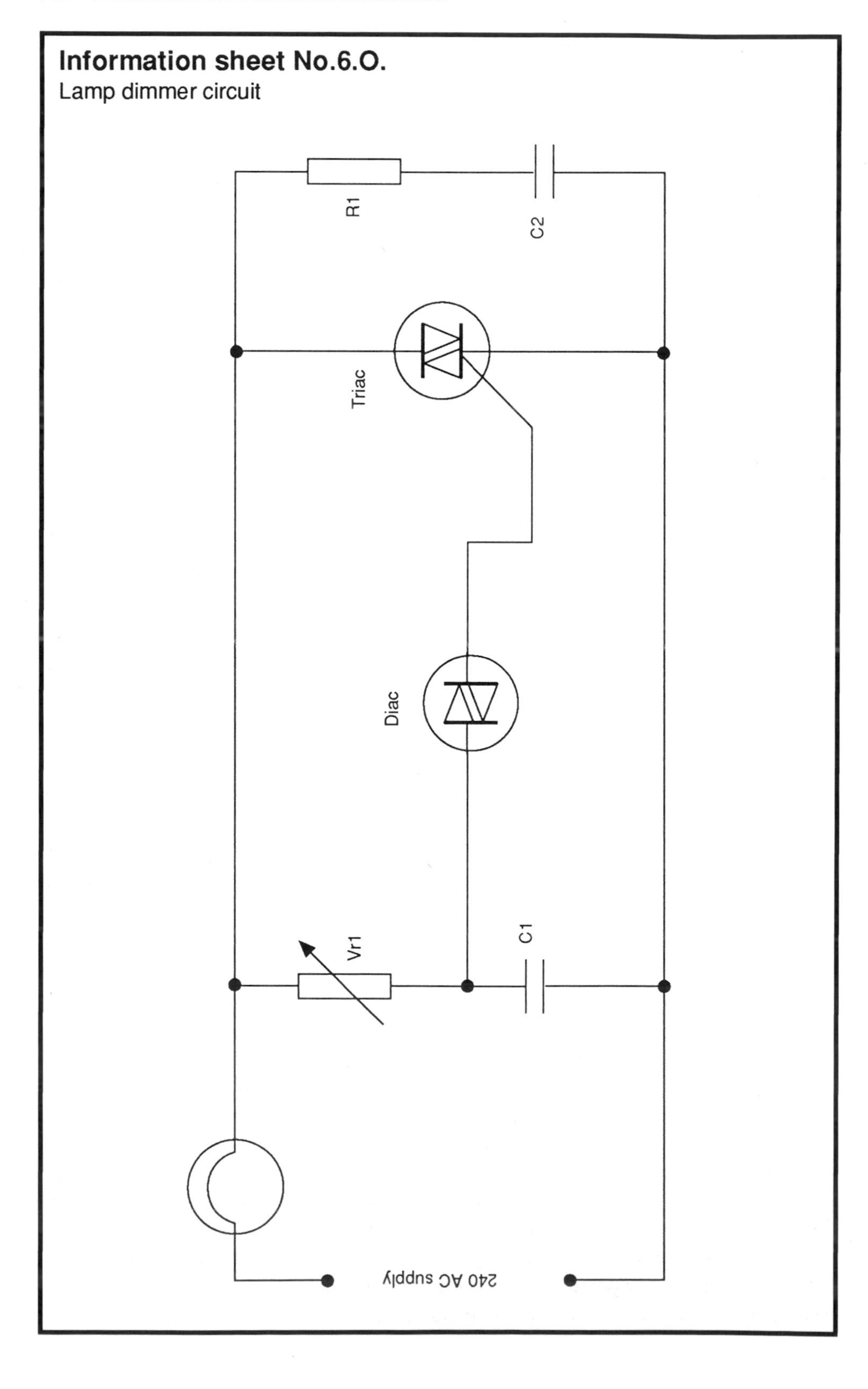

provide individual time control of domestic central heating systems and/or hot water supply, in conjunction with the appropriate thermostats (room/cylinder) and three port valve.

Diagram (b) is the internal circuit diagram for the programmer. It shows that the central heating and hot water electrical circuits are controlled by two independent changeover microswitches, the operation of which is determined by the programme settings. The timing for each operating period is set on a 24-hour dial using adjustable tappets which are usually colour-coded for simplification of selection. The selected programme for either circuit can be manually overridden at any time by operation of override switches. The programmer reverts back to the set programme automatically at the start of the next operating period.

The positioning of the above equipment in relation to the water system is shown in diagram (c).

6.7 Lamp dimmer

The circuit diagram shown on Information Sheet No. 60 is for a typical lamp dimmer system, using phase control; it incorporates the following units:

(1) Lamp;
(2) Variable resistor and capacitor for charging rate control;
(3) Diac and triac for the triac firing stage;
(4) 100 Ω resistor and 0.1 μF for the snubber network.

The time taken for the capacitor to charge is dependent upon the settings of the variable resistor. With this set to the minimum resistance, the capacitor will charge very quickly, and will continue to do so until it reaches the breakdown voltage of the diac. When this happens, breakover occurs, a pulse is sent to the gate of the triac, and current flows through the lamp until the triac turns off at the end of the positive half-cycle.

During the negative half-cycle the capacitor commences to charge again with the opposite polarity; since none of the components in the circuit are polarized this does not matter. Again, the triac is turned on when the breakdown voltage of the diac is reached, and current flows through the lamp.

With the variable resistance set to maximum the capacitor will charge slowly, so it takes longer for the diac to reach breakdown voltage; this means that the triac is switched on towards the end of each half cycle. The result of this is less current through the lamp, and the lamp is dimmer.

It will be seen that in parallel with the triac is a resistor–capacitor (R–C) series network; this is called a snubber network and prevents false triggering of the device by filtering out high voltage spikes in the mains.

Thus we can summarize by saying that the variable resistance controls the time taken to charge the capacitor, which in turn controls the point in time when the triac is turned on, which in turn determines the amount of current through the lamp. This method, which involves chopping the waveform, is known as phase control.

This sort of circuit works on the switching principle and not the resistive principle used by some other dimmers. Therefore, no power is wasted in its operation.

SUMMARY

This chapter has described electronic circuits and systems. The following important points should have been learned:

- That electricians must be able to recognize block diagrams and circuit diagrams when used to describe control systems for:
  (1) electric motor speed control;
  (2) security protection systems;
  (3) washing machine control circuits;
  (4) d.c. power supplies;
  (5) space heating systems;
  (6) lamp dimmer controls;
- How each of the units represented by a block on the block diagram interact with each other;
- That each of the blocks on a block diagram can represent either a unit (e.g. a thermostat, electric motor or a personal alarm button) or a stage of an electronic circuit such as the d.c. supply section, the amplifier or the filter;
- The principles of the open loop and the closed loop systems;
- The meaning of the expressions:
  (1) negative feedback;
  (2) positive feedback.

Test 6

Choose which one of the four answers is the correct one.

(1) In a closed loop system the output and input are:

(a) monitored,
(b) not monitored,
(c) wired in series,
(d) wired in parallel.

(2) Negative feedback in a system will:

(a) change the polarity of the circuit,
(b) blow the control circuit fuse,
(c) enhance the input conditions,
(d) oppose the input conditions.

(3) In a dimmer circuit the *snubber network*

(a) filters out high voltage spikes in the mains,
(b) alters the charging rate of the capacitor,
(c) consists of the diac and triac,
(d) consists of the variable resistor and capacitor.

(4) The main component in a simple half wave rectifier circuit is:

(a) a bridge rectifier,
(b) a tapped transformer,
(c) a single diode,
(d) a variable resistor.

(5) A *thermocouple* can be used in a space heating system as a:

(a) means of firing the boiler,
(b) flame failure device,
(c) room temperature controller,
(d) hot water temperature controller.

(6) The stabilizer stage of a power supply will:

(a) keep the input voltage stable,
(b) prevent the power unit from falling over,
(c) change the a.c. supply to d.c.,
(d) keep the output voltage stable.

(7) A speed-control error detector senses the difference between:

(a) negative feedback and positive feedback,
(b) the output of the tachogenerator and set speed control,
(c) the stabilized voltage and the nonstabilized voltage,
(d) a filtered supply and a non-filtered supply.

(8) The output of the tachogenerator is:

(a) always a.c.,
(b) indirectly proportional to the motor's speed,
(c) directly proportional to the motor's speed,
(d) always d.c.

(9) The arrows on the block diagram of a washing machine control:

(a) show an input signal from pump to controller,
(b) indicate an input signal from heater to controller,
(c) indicate an input signal from motor to controller,
(d) show an input signal from water level sensor to controller.

(10) The arrows on the block diagram of a space heating system:

(a) show output signals from controller to cylinder thermostat,
(b) indicate output signals from controller to pump,
(c) indicate input signals from boiler to controller,
(d) show input signals from fuel valve to controller.

Chapter 7
Calculations with Electronic Components

7.1 Calculations with resistors

Resistors in series

Several resistors may be connected in series, as shown in Fig. 7.1. The total resistance of the network is simply the sum of all the individual resistances, so that:

$$R_t = R_1 + R_2 + R_3$$

where R_t is the total resistance.

EXAMPLE 7.1

$$R_t = 10 + 22 + 33$$
$$= 65\ \Omega$$

Rule 1 for series circuits When resistors are connected in series the current must flow through each resistor in order to complete the circuit so that we can say: 'the current flowing in a series circuit is common to all the resistors in that circuit'.

Voltage drop in series circuits

Although the current is common to each resistor in the series circuit, as we have seen above, the voltage drop across each resistor will be different. The reason for

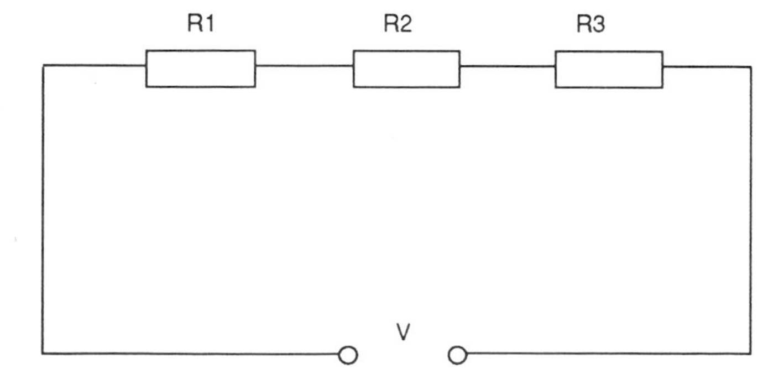

Fig. 7.1 Resistors in series.

this is quite simple: each resistor is of a different size, and since the circuit must obey Ohm's Law, i.e. $V = I \times R$, then, because each resistor is of a different value, each resistor will have a different voltage drop across it.

Let us take the three resistors used in the example above and see how this works.

EXAMPLE 7.2

Three resistors of values 10, 22, and 33 Ω respectively are connected in series to a 100 V d.c. supply. Calculate:

(1) The current flowing through the circuit;
(2) The voltage drop across each resistor.

Using Ohm's Law:

$$I = \frac{V}{R}$$

$$I = \frac{100}{65}$$

$$I = 1.53846 \text{ A}$$

Each individual voltage drop can now be calculated applying Ohm's Law:

$$\begin{aligned}
\text{Voltage drop across } R_1 &= I \times R = 1.53846 \times 10 = 15.3846 \\
\text{Voltage drop across } R_2 &= I \times R = 1.53846 \times 22 = 33.84612 \\
\text{Voltage drop across } R_3 &= I \times R = 1.53846 \times 33 = \underline{50.76918} \\
& \qquad\qquad\qquad\qquad\qquad\qquad\quad 99.99999
\end{aligned}$$

It will be no surprise for us to see that the individual voltage drops add up to the total applied voltage (give or take small inaccuracies).

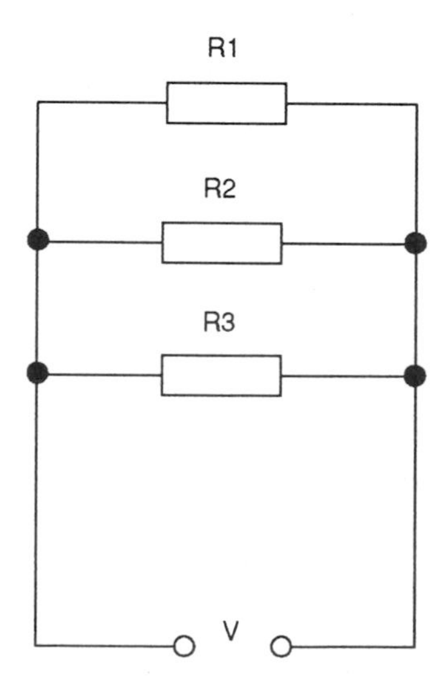

Fig. 7.2 Resistors in parallel.

Rule 2 for series circuits The voltage drop across each resistor will vary, and the sum of the individual voltage drops will add up to the total voltage.

Resistors in parallel

When resistors are connected in parallel, as shown in Fig. 7.2, then the total resistance of the circuit can be found from:

$$\frac{1}{R_t} = \frac{1}{R_1} + \frac{1}{R_2} + \frac{1}{R_3}$$

EXAMPLE 7.3

Three resistors of value 10, 20 and 30 Ω are connected in parallel. What is the total resistance of the circuit?

$$\frac{1}{R_t} = \frac{1}{10} + \frac{1}{20} + \frac{1}{30}$$

$$= \frac{6 + 3 + 2}{60}$$

$$= \frac{11}{60}$$

Inverting both sides gives:

$$R_t = \frac{60}{11}$$

$$R_t = 5.45\ \Omega$$

It will be found that the total resistance of the circuit is always less than the smallest resistance in the circuit.

Rule 1 for parallel circuits The supply voltage is common to all resistors.

The current in a parallel circuit

Although the voltage is common to each resistor in the parallel circuit, as we have seen above, the current through each resistor will be different. The reason for this is quite simple: each resistor is of a different size and since the circuit must obey Ohm's Law, i.e. $V = I \times R$, then, because each resistor is of a different value, each will have a different current flowing through it.

Let us take the example above and see how this works.

EXAMPLE 7.4

Three resistors of value 10, 20, and 30 Ω respectively are connected in parallel to a 20 V d.c. supply. Calculate:

(1) The total current flowing in the circuit;
(2) The current flowing through each resistor.

The total current can be found using Ohm's Law:

$$I = \frac{V}{R} = \frac{20}{5.45} = 3.66\ A$$

and each individual current through each resistor can be calculated using Ohm's Law:

$$\text{Current through } R_1 = \frac{V}{R} = \frac{20}{10} = 2.0\ \text{A}$$

$$\text{Current through } R_2 = \frac{V}{R} = \frac{20}{20} = 1.0\ \text{A}$$

$$\text{Current through } R_3 = \frac{V}{R} = \frac{20}{30} = 0.66\ \text{A}$$

It will be no surprise to find that if the individual current values are added they equal the total current.

Rule 2 for parallel circuits The current through each resistor will vary , and the sum of the individual current values will add up to the total current.

Calculating the tolerance of resistors

EXAMPLE 7.5

A resistor has the following colour bands:

First band, red; second band, violet; third band, orange; fourth band, gold. What is the value of the resistor, its tolerance and its highest and lowest possible values?

From the colour code table on Information Sheet No. 2C:

Red	= 2	The first digit is then	2
Violet	= 7	The second digit is then	7
Orange	= 3	The number of noughts is	3

This gives a value of 27 000 Ω.

The fourth band is gold; from the table gold is 5% so the value is 27 000 Ω ± 5%.

The highest value will be 27 000 Ω plus 5%, so that:

First find 5%

$$\frac{27\,000}{100} \times 5 = 1350$$

Highest value	27 000 + 1350 = 28 350
Lowest value	27 000 – 1350 = 25 650

EXAMPLE 7.6

A resistor has the following colour bands:

Brown, red and black. What is its value, tolerance and highest and lowest values?

Brown = 1
Black = 0
Red = 00

This gives a value of 1000 Ω

The fourth band is absent, so a tolerance of 20% is assumed. The highest and lowest values are therefore:

First find 20%

$$\frac{1000}{100} \times 20 = 200$$

Highest value	1000 + 200 = 1200
Lowest value	1000 – 200 = 80

7.2 Calculations with capacitors

The value of capacitors is expressed in farads. These are rather large units so it is more usual in electronics to rate capacitors in fractions of a farad, so that:

one millionth	= a microfarad	= 1 μF	$= 1 \times 10^{-6}$ F
one thousand millionth	= a nanofarad	= 1 nF	$= 1 \times 10^{-9}$ F
one million millionth	= a picofarad	= 1 pF	$= 1 \times 10^{-12}$ F

We have seen that the ability of the component to store an electric charge is called its *capacitance*. By definition, a capacitor has a capacitance of one farad when a PD of one V maintains a charge of one coulomb on that capacitor, so that:

$$C = \frac{Q}{V}\ \text{(F)}$$

or

$$Q = CV$$

where: Q = charge (a coulomb is an ampere/second or It), C = capacitance (measured in farads or parts of a farad), and V = voltage (working voltage).

EXAMPLE 7.7

Calculate:

(1) The total charge on a 100 μF capacitor when exposed to a constant flow of 4 mA for 5 seconds;
(2) the voltage across the capacitor.

$$Q = I \times t = 4 \times 10^{-3} \times 5 = 0.02 \text{ coulombs}$$

and

$$Q = CV$$

so that by transposing

$$V = \frac{Q}{C} = \frac{0.02}{100 \times 10^{-6}} = 200 \text{ V}$$

Capacitors in series

Several capacitors can be connected in series as shown in Fig. 7.3. The total capacitance of the network may be found from the formula:

$$\frac{1}{C_t} = \frac{1}{C_1} + \frac{1}{C_2} + \frac{1}{C_3}$$

where C_t is the total capacitance.

EXAMPLE 7.8

Three capacitors of 2, 4 and 8 μF are connected in series. What is the total capacitance of the arrangement?

Now:

$$\frac{1}{C_t} = \frac{1}{2} + \frac{1}{4} + \frac{1}{8}$$

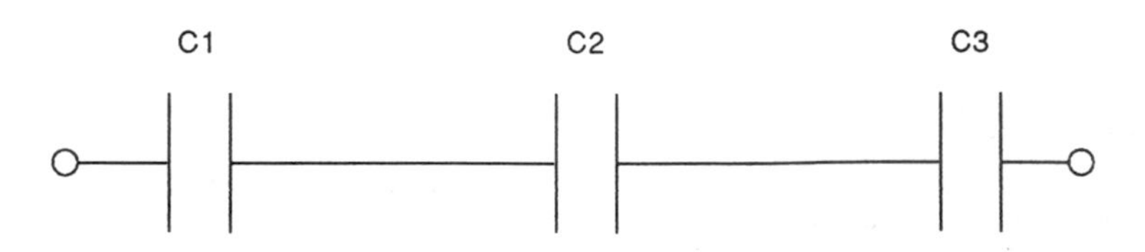

Fig. 7.3 Capacitors in series.

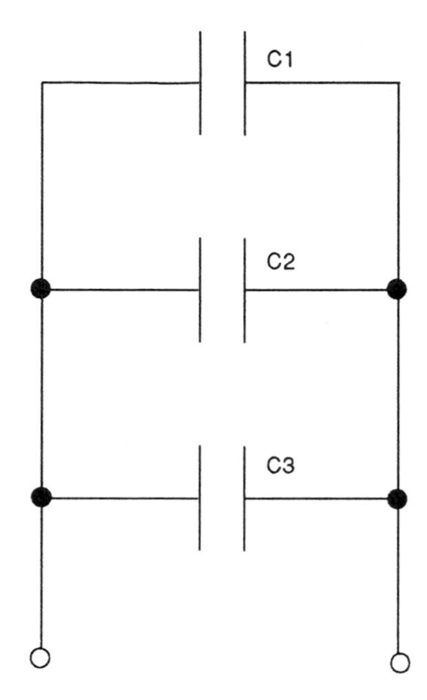

Fig. 7.4 Capacitors in parallel.

The lowest common denominator (LCD) is 8; therefore:

$$\frac{1}{C_t} = \frac{4 + 2 + 1}{8}$$

$$\frac{1}{C_t} = \frac{7}{8}$$

Turning both sides upside-down:

$$C_t = \frac{8}{7}$$

C_t = 1.142 μF total capacitance.

Capacitors in parallel

Several capacitors may be connected in parallel, as shown in Fig. 7.4. The total capacitance in this instance is the sum of all the values:

$$C_t = C_1 + C_2 + C_3$$

EXAMPLE 7.9

Three capacitors of value 2, 4, and 8 μF are connected in parallel. What is the total capacitance of the circuit?

$$C_t = 2 + 4 + 8$$

Therefore:

$$C_t = 14\ \mu\text{F total capacitance.}$$

Capacitors in alternating current circuits

When a capacitor is used in a circuit using only d.c. waveforms, and because in a d.c. supply the current flows only in one direction, the capacitor will charge up and maintain its charge (assuming it was a perfect device) even when the supply is disconnected. It will only release the charge when some external circuit provides a path for the discharge to take place.

When a capacitor is used in a circuit using only a.c. waveforms, the current flows in one direction for the first half of the cycle and then in the opposite direction for the second half of the cycle. For this reason, the capacitor charges up one way and then discharges as the wave reaches zero, then charges up the other way and discharges as the wave reaches zero again. In a.c. circuits, then, the capacitor never becomes charged in any direction.

As the frequency of the a.c. is increased, so more charging and discharging takes place. This action produces a phenomenon called capacitive reactance and the higher the frequency of the a.c. waveform, the lower the capacitive reactance. The symbol for reactance is X and capacitive reactance thus has the symbol X_C. Like resistance, capacitive reactance is measured in ohms. This reactance can be found by using the formula:

$$X_C = \frac{1}{2\pi fC}$$

and, because it is likely to be in microfarads, the formula becomes:

$$X_C = \frac{10^6}{2\pi fC}$$

where C is the value of the capacitor in farads (F), f is the frequency of the supply in hertz (Hz), 2π (2×3.142) is a constant, and X_C is the capacitive reactance in ohms.

EXAMPLE 7.10

A capacitor of value 22 μF is connected in turn across:

(1) an a.c. supply with a frequency of 1.5 kHz;
(2) an a.c. supply of 50 Hz frequency.

What would its capacitive reactance be under both conditions?

At 1.5 kHz it would be found from:

$$X_C = \frac{10^6}{2 \times \pi \times 1500 \times 22}$$

$$= 4.82\ \Omega$$

At 50 Hz it would be found from:

$$X_C = \frac{10^6}{2 \times \pi \times 50 \times 22}$$

$$= 144.68\ \Omega$$

7.3 Calculations with inductors

The value of an inductor is expressed in units called henries (H). The henry is a rather large unit and so it is more usual in electronics to rate inductors in fractions of a henry, so that:

one millionth	= a microhenry	= 1 μH	= 1×10^{-6} H
one thousand millionth	= a nanohenry	= 1 nH	= 1×10^{-9} H
one million millionth	= a picohenry	= 1 pH	= 1×10^{-12} H

We have seen in an earlier section (Section 2.14) that the ability of this component to resist the flow of a.c. is called its inductance (L). It is unlikely that the student will ever have recourse to connecting multiple banks of inductors in circuit. However, it is worth noting that if there is no magnetic coupling between the individual components then calculations are much like those for resistors, so that:

For inductors in series:

$$L_t = L_1 + L_2 + L_3 \ldots$$

and for inductors in parallel:

$$\frac{1}{L_t} = \frac{1}{L_1} + \frac{1}{L_2} + \frac{1}{L_3} \ldots$$

The above expressions do not hold good if the components are mounted in close proximity and mutual coupling takes place. If we consider two inductors in circuit where mutual inductance is taking place, the magnetic field produced by the current in the coil of the first inductor influences the field produced by the coil of the second inductor, and vice versa. This mutual inductance must be allowed for in the calculations but is beyond the scope of students reading this book.

The inductance of a coil is dependent on two factors:

(1) The number of turns on the coil;
(2) The material the core is made from.

When the inductor has a simple air core, then the inductance is proportional to the number of turns squared, i.e.:

$$L = n^2$$

Inductors in alternating current circuits

When an inductor is used in a circuit using only d.c. waveforms, and because in a d.c. supply the current flows only in one direction, a magnetic field is set up around the the coil of the inductor. Because it is a direct current this flux builds up to a maximum value, at which it remains until the supply is disconnected. On disconnection, the field collapses inwards again, inducing a back emf into the circuit.

When an inductor is used in a circuit using only a.c. waveforms, the current flows in one direction for the first half of the cycle and then in the opposite direction for the second half of the cycle. For this reason the field around the coil builds up one way and then collapses as the current falls to zero, then builds up the other way and collapses as the wave returns to zero, following the alternating current cycle.

The result of this flux growth and decay under a.c. conditions is to introduce into the circuit a phenomenon called inductive reactance. Inductive reactance, like resistance, is measured in ohms. The symbol for reactance is X and inductive reactance therefore has the symbol X_L. Like resistance, inductive reactance is measured in ohms. This reactance can be found by using the formula:

$$X_L = 2\pi fL$$

where L is the inductance of the coil in henries (H), f is the frequency of the supply in hertz (Hz), 2π (2×3.142) is a constant, and X_L is the inductance of the coil in ohms (Ω).

EXAMPLE 7.11

A coil has an inductance of 0.2 H and a supply current of frequency 50 Hz passes through it. What will be the inductive reactance of the coil?

Now:

$$\begin{aligned} X_L &= 2\pi fL \\ &= 2 \times 3.142 \times 50 \times 0.2 \\ &= 62.8\ \Omega \end{aligned}$$

7.4 Calculations with transformers

The rating of transformers is usually stated in kilovolt-amps (kVA) or in the case of electronic devices volt-amps (VA). Most transformers of whatever type have some losses, but in the perfect transformer the input to the primary would equal the output from the secondary, i.e.:

$$V_p \times I_p = V_s \times I_s$$

Transposing this we get:

$$\frac{V_p}{V_s} = \frac{I_s}{I_p} \quad \textbf{(7.1)}$$

In Chapter 2 we learned that the ratio of the input voltage to the output voltage depended on the ratio of the number of turns on the primary winding to the number of turns on the secondary winding, and this can also be expressed as:

$$\frac{V_p}{V_s} = \frac{N_p}{N_s} \quad \textbf{(7.2)}$$

Combining equations (**7.1**) and (**7.2**) we get

$$\frac{V_p}{V_s} = \frac{N_p}{N_s} = \frac{I_s}{I_p}$$

Typical problem with transformers

A 240 V, 50 Hz, transformer which is part of a fire-alarm system burns out and we are required to replace it. The original manufacturer is no longer trading and it is decided that we will replace it with one from a well known supplier of radio spares who stocks a suitable range of transformers from 20 to 200 VA rating. The replacement transformer requires the secondary winding to be wound to our own specification, and the engineer in charge asks you to find out the requirements.

On carrying out further checks we find the following:

- A check on the fire alarm system shows that the secondary voltage required is 24 V and that the secondary current demand for the system is in the region of 7.5 A;
- From the catalogue, the recommended number of turns of 1.5 mm dia. insulated copper wire; for the secondary winding of this range of transformers it is 28 turns for the 20 VA, 40 turns for the 50 VA, 60 turns for the 100 VA and 99 turns for the 200 VA.

From the above information we are to determine the following:

(1) Which of the VA ratings would be suitable for our purpose;
(2) What would be the maximum current output;
(3) How many turns would be on the primary winding;
(4) How much current it will draw from the mains.

Answering (1) on page 157, we know that the V_s is 24 V and that the current demand of the equipment is 7.5 A. The actual required VA can be found from:

$$\begin{aligned} VA &= V_s \times I_s \\ &= 24 \times 7.5 \\ &= 180\ VA \end{aligned}$$

The 180 VA represents the actual VA required, so that the 200 VA rated transformer will be the one to use.

Answering (2) on page 157, the transformer will have a rating of 200 VA and as the primary voltage (V_p) is to be 24 V the maximum current from the output side of the transformer (I_s) can be found from:

$$VA = V_s \times I_s$$

Transposing for I_s:

$$I_s = \frac{VA}{V_s}$$

$$I_s = \frac{200}{24}$$

$$I_s = 8.3333\ A\ \text{maximum.}$$

The system requires 7.5 A so this is within the maximum. Answering (3) on page 157, this is of academic interest only as the transformers in this range have the primary pre-wound. However, the number of primary windings (N_p) can be found from:

$$\frac{V_p}{V_s} = \frac{N_p}{N_s}$$

so that:

$$N_p = \frac{N_s \times V_p}{V_s}$$

Therefore:

$$N_p = \frac{99 \times 240}{24}$$

$$= 990\ \text{turns.}$$

Answering (4) on page 157, the current taken from the mains (I_p) can be found from:

$$\frac{V_p}{V_s} = \frac{I_s}{I_p}$$

Transposing for I_p:

$$I_p = \frac{I_s \times V_s}{V_p}$$

$$I_p = \frac{8.3333 \times 24}{240}$$

$$I_p = 0.83333 \text{ A.}$$

We would order a 200 VA transformer and have the secondary wound with 99 turns of 1.5 mm dia. insulated wire. After fitting the replacement transformer, check the protection as this should have operated and saved the transformer from burning out. From the above, a BS 4265 1 A anti-surge cartridge fuse would suffice. These are specially designed for the protection of transformers and motors. Power supplies are discussed more fully in Chapter 6 of this book.

The above calculations could have been carried out using ratios, as explained in Chapter 2. Once we have a pair of components of the above compound equation we can work out the ratio and then use this to find the others' values.

EXAMPLE 7.12

$$\frac{V_p}{V_s} = \frac{240}{24} = 10$$

Our transformer has a ratio of 10:1.

If we know the primary turns (N_p) are 99, then using the 10:1 ratio, the secondary turns (N_s) must be 990, and our calculations above show this is correct. By the same token if our secondary current is calculated at 8.3333 A, then the primary current (I_p) must be 0.83333. A look at the calculations will show that this, too, is correct.

SUMMARY

This chapter has described calculations involving electronic components. The following important points should have been learned:

- That electricians must be able to carry out simple calculations involving electronic components;
- How to carry out calculations with resistors, capacitors and inductors connected in both series and parallel;
- How to carry out calculations with resistors, capacitors and inductors connected in a.c. circuits;

- How to determine the value of resistors and calculate the tolerance range;
- How to determine the value of capacitors and calculate the charge on a capacitor.

Test 7

Choose which one of the four answers is the correct one.

(1) Resistors of value 20 and 10 Ω are connected in parallel; the total resistance of the circuit (in ohms) is:

(a) 30,
(b) 200,
(c) 6.6,
(d) 15.

(2) Resistors of value 20 and 10 Ω are connected in series; the total resistance of the circuit (in ohms) is:

(a) 30,
(b) 200,
(c) 6.6,
(d) 15.

(3) Capacitors of value 2, 4 and 8 μF respectively are connected in parallel; the total capacitance (in microfarads) is:

(a) 14.00,
(b) 0.14,
(c) 0.014,
(d) 1.14.

(4) Capacitors of value 2, 4 and 8 μF respectively are connected in series; the total capacitance (in microfarads) is:

(a) 1.400,
(b) 14.00,
(c) 114.30,
(d) 1.143.

(5) An 80 μF capacitor is connected across a 500 V d.c. supply; the charge (in coulombs) will be:

(a) 4.00,
(b) 0.04,
(c) 8.00,
(d) 0.08.

(6) A 240 V/24 V transformer has 50 secondary turns; therefore, on its primary, it will have:

(a) 200 turns,

(b) 300 turns,
(c) 400 turns,
(d) 500 turns.

(7) A purely inductive circuit of 0.1 H is connected to a 100 V 50 Hz supply; the inductive reactance of the circuit (in ohms) will be:

(a) 31.40,
(b) 3.14,
(c) 500,
(d) 5.

(8) A purely capacitive circuit of 30 μF is connected to a 100 V 50 Hz supply; the capacitive reactance of the circuit (in ohms) will be:

(a) 15 000,
(b) 1500,
(c) 10.62,
(d) 106.20.

Chapter 8
Waveforms

8.1 Types of waveform

We have seen that in electronic circuits many different voltages and currents exist in the various parts of the circuit. The different signals carried by these voltages and currents also vary, and it is often useful to our understanding of the different concepts involved if we can consider these variations in graphical form.

A graph is a useful way of doing this, as we can depict the magnitude of these electrical quantities as functions of time. The shape that the graph takes up is known as the *waveform* of the voltage or current being considered, and if it repeats itself at regular intervals, it is termed a periodic waveform.

Valuable information about a circuit can be obtained by careful study of the waveforms produced and indeed we shall see later when we discuss oscilloscopes that some faults cannot be detected by other means. Information Sheet No. 8A shows some of the more common waveforms found in electronic applications; these are as follows:

(1) *Sine waves* – produced by the a.c. mains supply and the harmonic type oscillator;
(2) *Square waves* – produced by the relaxation type oscillator; uses include timing and counting circuits;
(3) *Sawtooth waves* – for the time base of the cathode ray oscilloscope which we shall be discussing later;
(4) *Pulse waves* – used to trigger thyristors or triacs and also in digital circuitry.

Sine waves

The mains supply provided by the power companies in this country is a.c. rather than d.c. This is because a.c. can be stepped up by the use of transformers to very high voltages for the efficient transmission of the electricity and stepped down again where it is required to be used.

We saw in Chapter 1 how the movement of a conductor in a magnetic field could induce a voltage in that conductor and if the conductor was arranged in the form of a coil a greater voltage would be generated. To produce the a.c. supply discussed above, we would require an arrangement as shown on Information Sheet No. 8B. The waveform produced by this arrangement would be sinusoidal and the sine wave it produces is shown in Fig. 8.1. A full description of the

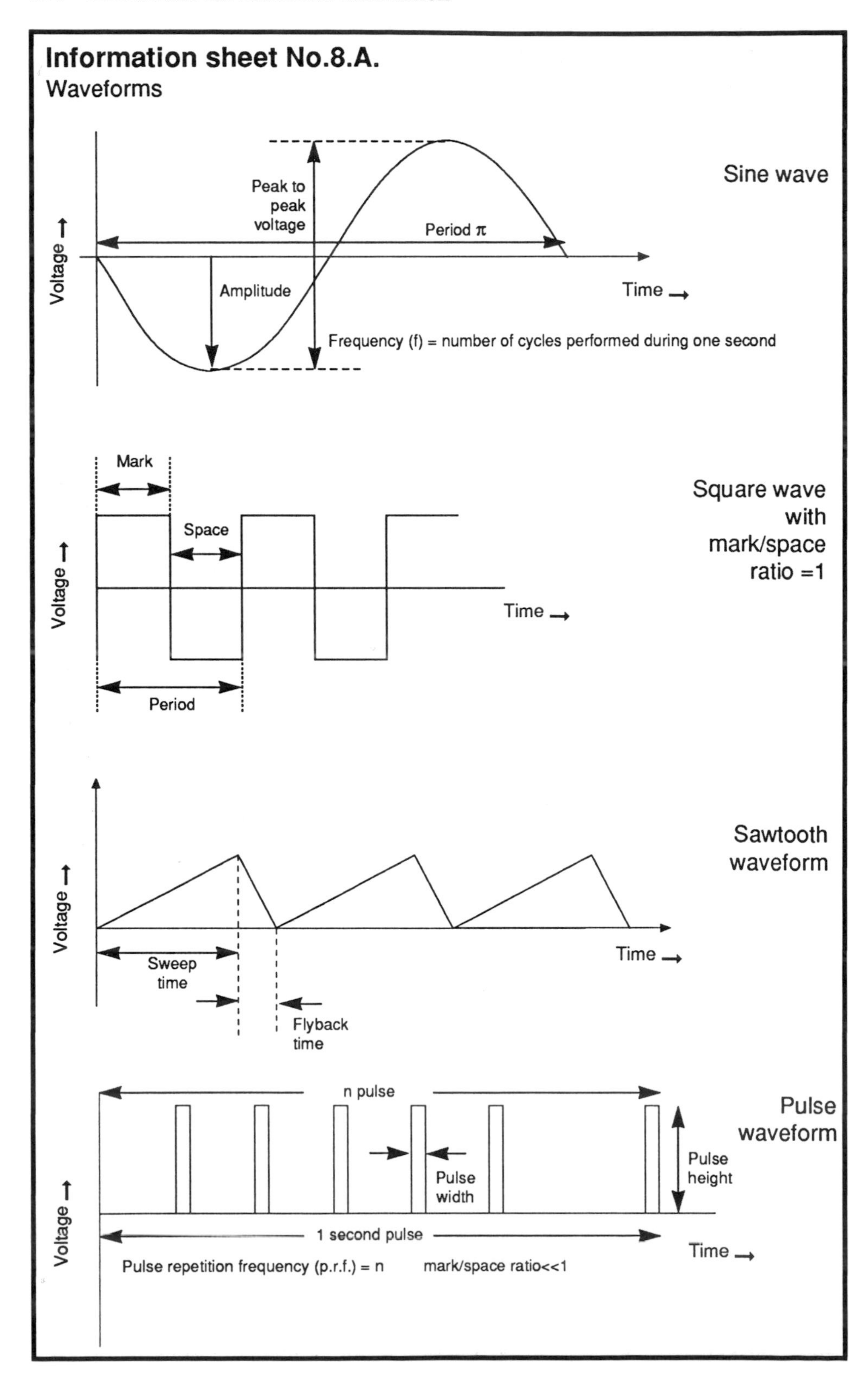

Information sheet No.8.A.
Waveforms
Sine wave
Peak to peak voltage
Period π
Voltage
Amplitude
Time
Frequency (f) = number of cycles performed during one second
Mark
Square wave with mark/space ratio =1
Space
Voltage
Time
Period
Sawtooth waveform
Voltage
Time
Sweep time
Flyback time
n pulse
Pulse waveform
Pulse width
Pulse height
Voltage
1 second pulse
Time
Pulse repetition frequency (p.r.f.) = n
mark/space ratio<<1

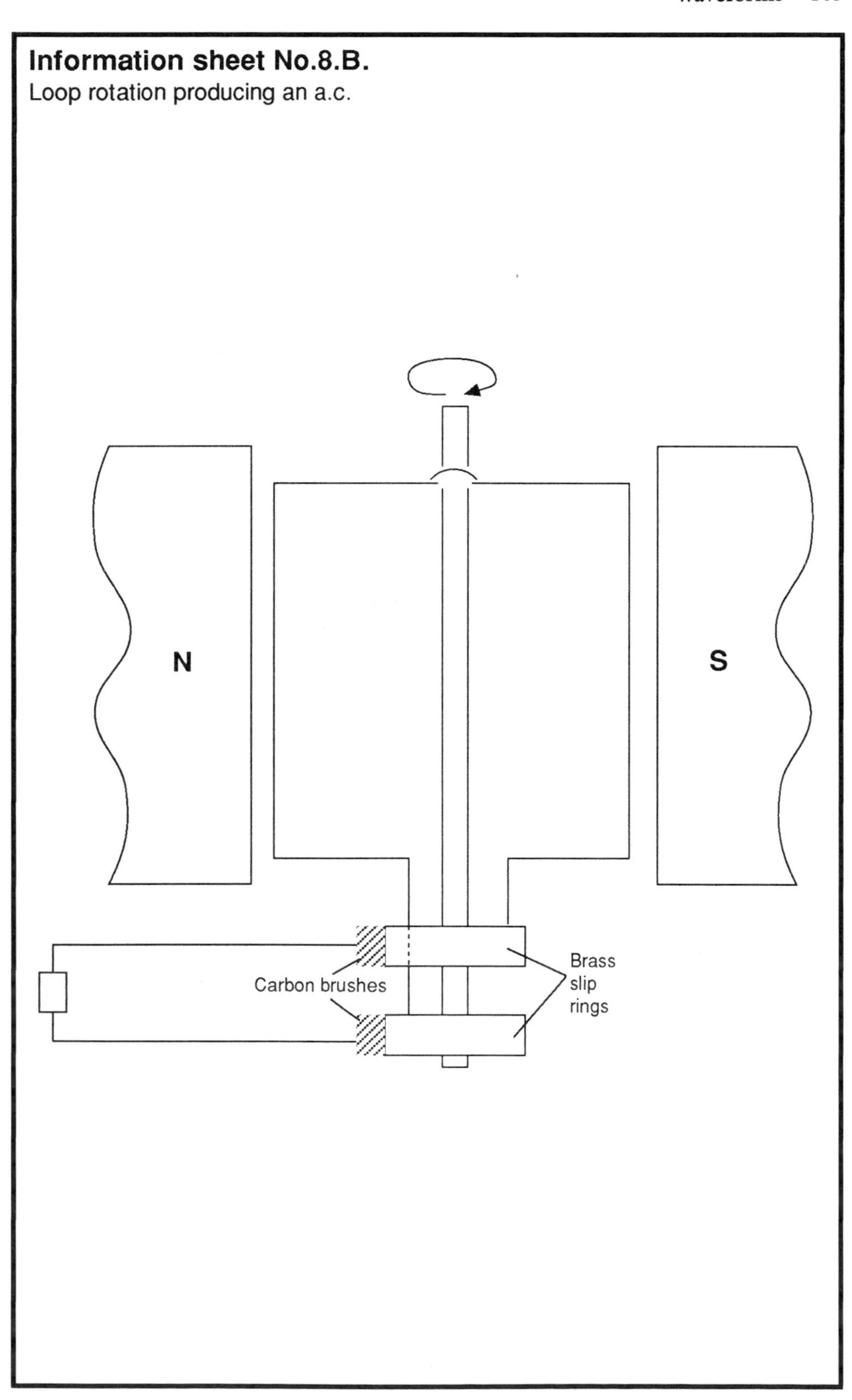
Information sheet No.8.B.
Loop rotation producing an a.c.
N
S
Carbon brushes
Brass
slip
rings

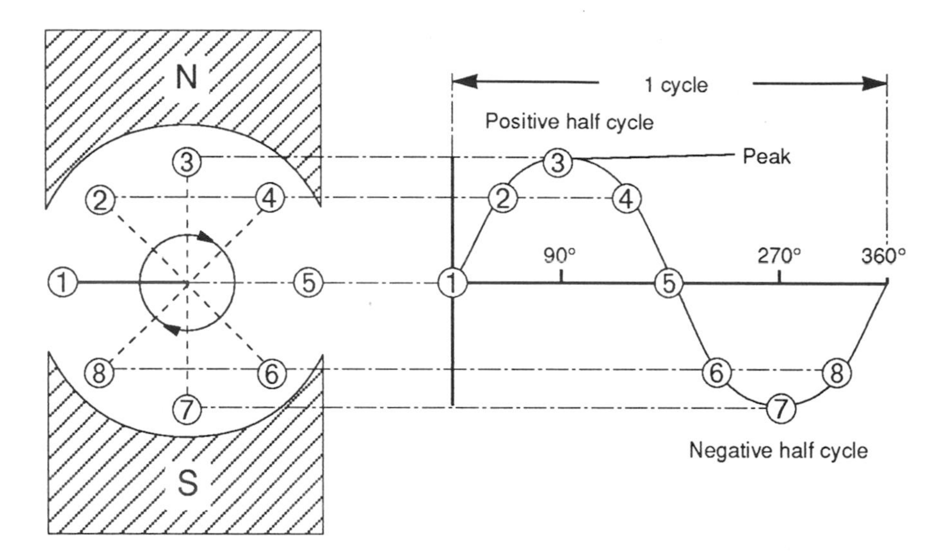

Fig. 8.1 A.C. sine wave.

generation of a.c. supplies is given in Book Three of *Electrical Installation Practice*.

In electronics the a.c. supply is usually rectified to d.c. However, if we require voltages that are sinusoidal with time, then an oscillator is used. An oscillator is an electronic circuit which, in the broadest definition, converts d.c. to a.c. We could, however, think of it also as an electronic circuit that generates waveforms. Two main forms of oscillator exist:

- The harmonic type, which generates sine waves; and
- The relaxation type, which generates square, sawtooth and pulse waves.

Examples of these waveforms are shown on Information Sheet No. 8A.

Square waves

One transistor circuit which will be of interest to us because it produces a square waveform is the *multivibrator circuit* shown in Fig. 8.2; it is of the astable or free running type. Technically speaking, it is a relaxation oscillator which uses the principle of commutation (switching).

The circuit is a sequential circuit, i.e. a circuit that goes through a series of changes one after the other. The multivibrator has two distinct states, and it changes, or is forced to change, from one state to the other. If we can imagine the circuit as changing or vibrating from one state to the other over and over again we can understand where it got its name.

The multivibrator oscillates at a frequency dependent on the values of C1, C2, R2 and R3, and the cross-coupling and high mutual amplification in the circuit causes the transistors to be turned on and off alternately. The output voltage at the

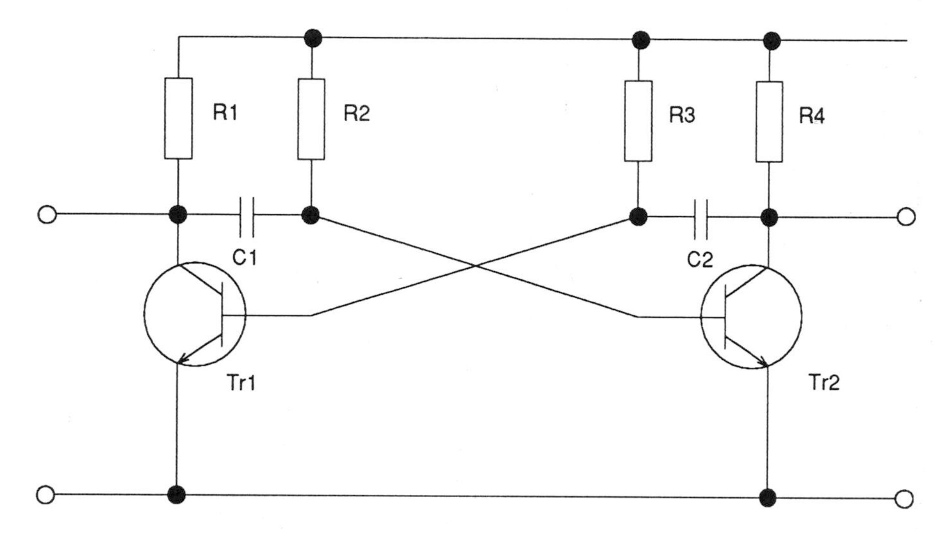

Fig. 8.2 Astable multivibrator.

collectors of each of the transistors will alternate from high (approximately +8 V) to low (almost 0 V). This will produce in theory a series of square or rectangular waveforms, but in practice these are not perfectly square due to the delay when the switching takes place.

Sawtooth waves

The majority of signals studied with an oscilloscope are periodic, that is to say they are repeated exactly in equal successive periods of time. All oscilloscopes, therefore, include a special type of relaxation oscillator and this, together with the X-amplifier circuit, constitute what is known as a time base. The function of the

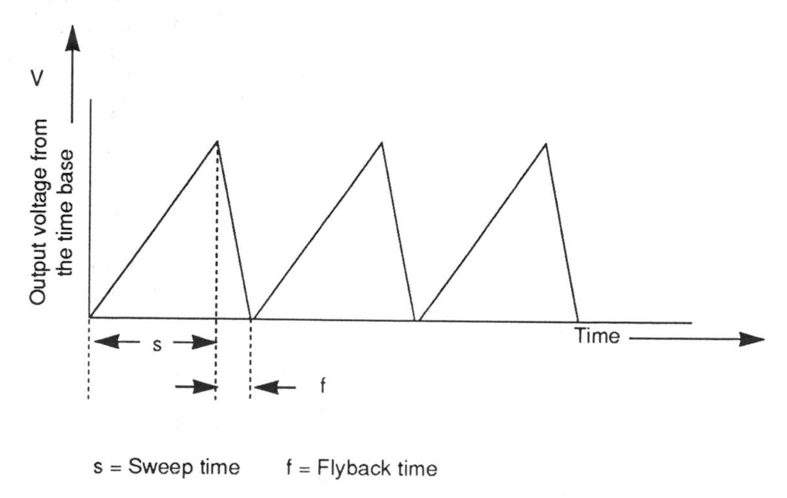

Fig. 8.3 Sawtooth waveform.

time base is to deflect the beam of the oscilloscope in the X-direction so that the spot moves from left to right across the screen with uniform speed. Having moved across the screen, it must return to its original position in the shortest time possible. This requirement is met by the sawtooth waveform whose characteristics (a long rise time and quick discharge time) match exactly the long sweep time and a short flyback time required (see Fig. 8.3).

Pulse waves

We learned in our earlier studies that digital integrated circuits using combinational logic, use signals that are either on or off. Circuits generating pulse waves can be built from discrete components, but nowadays one of the integrated circuits such as the *555 timer* would be utilized. Essentially, the circuit is simple as only three additional external components are required to construct a highly efficient pulse generator (see Fig. 8.4).

The circuit is not sensitive to temperature change and is independent of the supply voltage, which may be between 3 and 18 V. The period of time during which the wave is high (logic 1) and the time that it is low (logic 0) are determined by the choice of resistors and capacitors; however, the *on* time must exceed the *off* time.

A CMOS version of the 555 timer, generally known as the ICM 7555, is available and can usually be used as a direct replacement. This has the advantage of much lower power consumption and with the smaller value resistors will

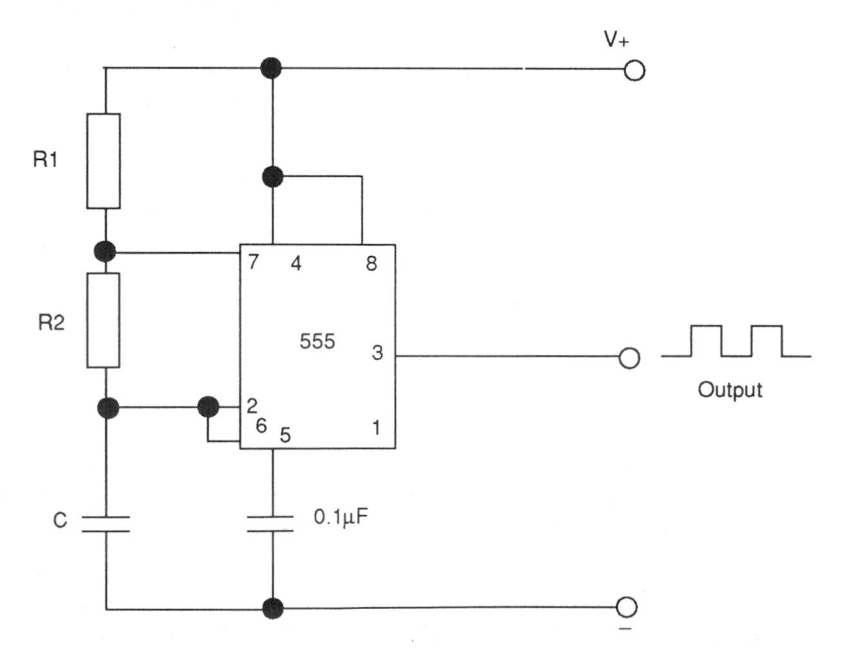

Fig. 8.4 555 pulse generator circuit.

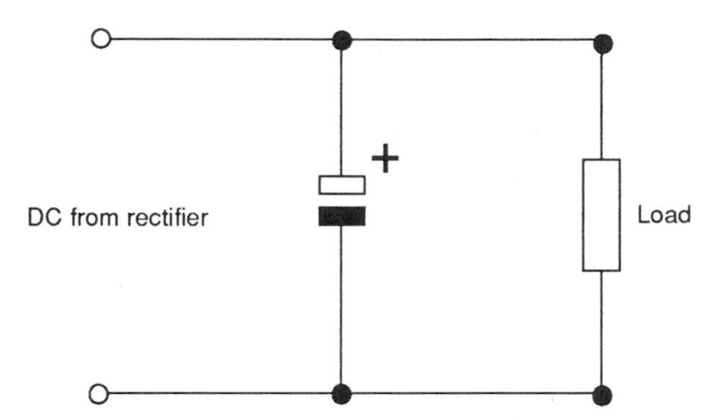

Fig. 8.5 Smoothing circuit.

generate a *high* pulse of precisely fixed length but only a few milliseconds or microseconds long.

8.2 Smoothing

As stated earlier (Section 8.1), the a.c. supply is usually rectified to d.c. for use in electronic circuits and rectification is discussed in detail in Chapter 6. However, most electronic circuits would not function correctly with a raw d.c. supply and therefore the output from a bridge rectifier must be modified further to give a *smooth* d.c. supply. The usual refinement is to fit a large electrolytic capacitor across the output, and so in parallel with the load. This has the effect of smoothing the output, as, when the waveform is rising from zero to maximum, the capacitor will charge up. When the waveform starts to fall to zero, the capacitor will start to discharge and so feed the output. Generally speaking, the larger the smoothing capacitor, the smoother the output. The output voltage on-load of the smooth d.c. supply is shown in Fig. 8.5.

Students continually ask: 'Can the smoothing capacitor be too large?' The answer in theory is 'no', but in practice, at *switch-on*, a really large capacitor will have a high inrush current; this will probably continually blow the secondary fuses. This problem may be overcome by fitting anti-surge fuses.

8.3 Ripple

Not all power supplies provide a completely smooth d.c. output. It is possible that a small deviation in the d.c. output may be present; this is called ripple. Ripple is the difference between the peak and the trough of the output waveform. This is clarified in Fig. 8.6.

The size of the smoothing capacitor will depend largely upon the current taken by the load. For a load current of 100 mA, then, a 220 μF should be adequate, rising to 4700 μF for 1 A, and a massive 47 000 μF for a 10 A load current.

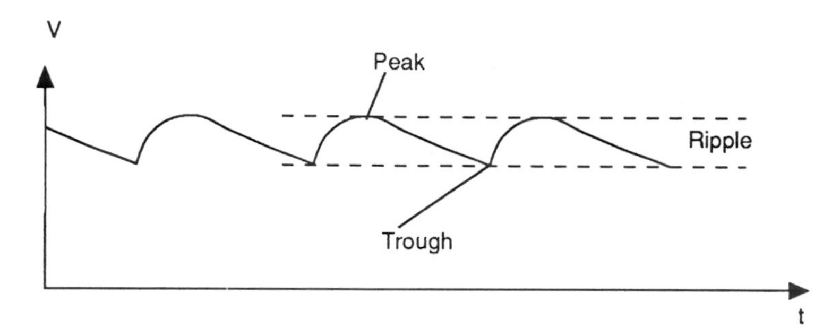

Fig. 8.6 Example of ripple in d.c. supply.

So far, we have come to realize that the alternating high voltage of the mains supply may be rectified to provide a low voltage smooth d.c. supply for the load. There are many applications where this is not enough. In electrically noisy environments it is likely that mains-borne high voltage high frequency spikes (sometimes called transients) may be present on the supply. To prevent these spikes being transferred to our power supply a low pass filter is included in circuit.

8.4 Filters

A filter is an electronic circuit that is designed to respond to a particular frequency or band of frequencies. The filter will *let through* signals of a particular selected range, but signals outside this range will be *blocked* or attenuated. It follows that a filter is frequency sensitive and there are many examples in electronics where this can be used to our advantage.

Filter circuits using reactive components such as inductors, capacitors, or both, are known as passive filters. Ideally these should be pure inductors and pure capacitors; realistically they will possess some resistance. Circuits using semiconductor components such as an op-amp (see Chapter 4), as you would expect, are called active filters.

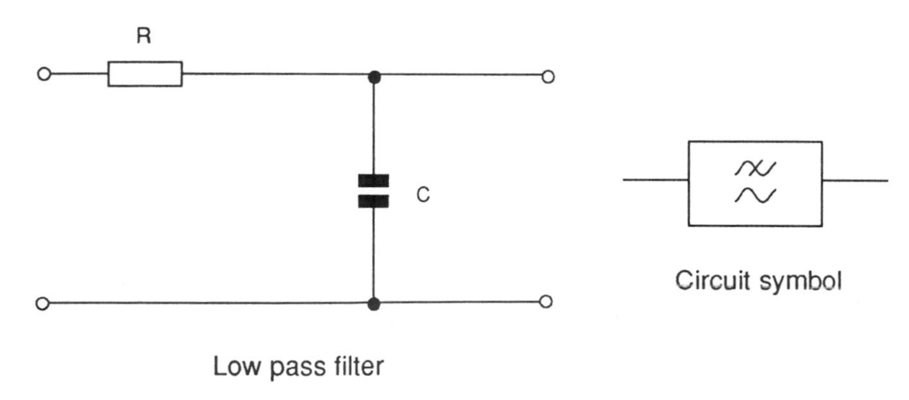

Fig. 8.7 Low pass filter circuit and symbol.

The filter circuit is used in many types of electronic circuit but particularly in communications - radio transmitters and receivers - and also power supply circuits, as we have seen in our earlier work. The frequencies with which they work vary from megahertz in the case of communication circuits to a few hertz in the case of power supplies. Four basic categories of filter exist:

- Low pass,
- High pass,
- Band pass,
- Band stop.

Low pass filters

These are designed to pass all signals below a certain frequency, say, from direct current to 100 hertz (the cut-off point), and give high attenuation above this value. They are often used to block a.c. ripple on d.c. power supplies; a simple low pass filter circuit consisting of a capacitor and a resistor is shown in Fig. 8.7.

The principle of operation is that at low frequencies the capacitive reactance of a capacitor is high (and with d.c. infinite!); therefore, this circuit will allow low frequencies to pass more easily than high frequencies.

High pass filters

These are designed to pass all signals above a certain frequency, and attenuate signals below the cut-off frequency. They are often used to to block d.c. when a number of a.c. amplifier stages are connected together; a simple high pass filter circuit consisting of a resistor and capacitor is shown in Fig. 8.8.

The circuit is very similar to the low pass filter except that the resistor and capacitor positions are exchanged. Again, this circuit works by the variation of the

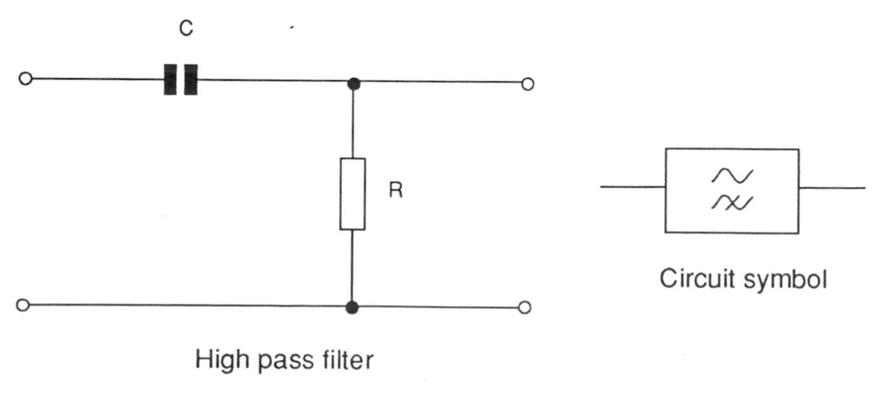

Fig. 8.8 High pass filter circuit and symbol.

Information sheet No.8.C.
Further filters

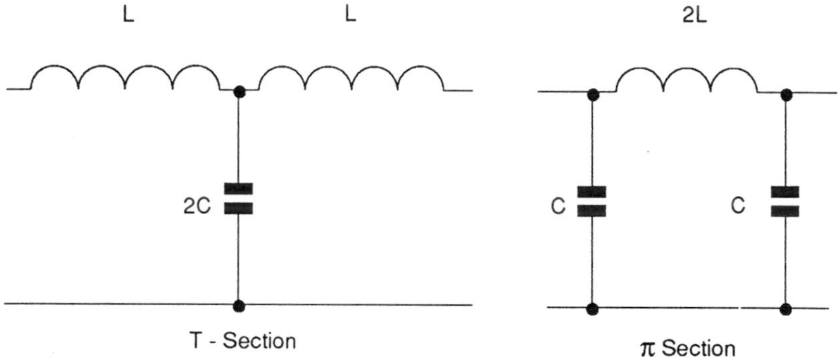

L L 2L

2C C C

T - Section

π Section

Low pass

C C C/2

L/2 L L

T - Section

π Section

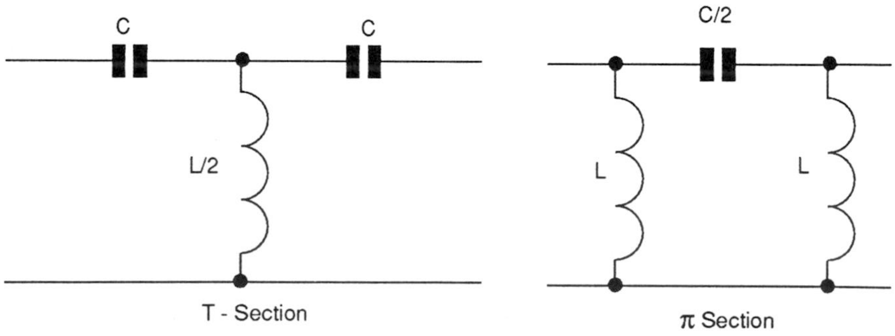

High pass

capacitor's reactance with frequency. In this case at low frequencies the reactance will be high and so block the signal, whereas at high frequencies the reactance will be low and so pass the signal.

Band pass filters

Band pass filters, as their name suggests, are designed to pass all frequencies within a limited frequency band or range, say 1 kHz to 2 kHz, attenuating all signals outside this range.

In its simplest form this filter consists of an inductor and a capacitor connected in series.

Band stop filters

These are designed to block or attenuate signals in a particular frequency band or range.

In its simplest form this filter consists of an inductor and a capacitor connected in parallel.

Further examples of the filter circuits mentioned above are shown on Information Sheet No. 8C.

SUMMARY

This chapter has described waveforms associated with electronic circuitry. The following important points should have been learned:

- That electricians must be able to recognize the following waveforms:
  (1) Sine wave;
  (2) Square wave;
  (3) Sawtooth wave;
  (4) Pulse wave.
- In simple terms, which circuit produces which type of waveform;
- The meaning of the terms *smoothing*, *ripple* and *filter*. The student should also have a simple concept of the components and circuits producing these effects.

Test 8

Choose which of the four answers is the correct one.

(1) Ripple on an output waveform is the difference between:

(a) the a.c. and the d.c. voltages,
(b) the peak and the trough,
(c) input frequency and output frequency,
(d) average and RMS voltage.

(2) A multivibrator circuit produces a waveform that is:

(a) square,
(b) triangular,
(c) sawtoothed,
(d) circular.

(3) Mains-borne high-voltage high frequency spikes are eliminated or reduced by a suitable:

(a) centre-tapped transformer,
(b) bridge rectifier,
(c) low pass filter circuit,
(d) high pass filter circuit.

(4) The waveform produced by an a.c. generator takes the form of a:

(a) square wave,
(b) sawtooth wave,
(c) triangular wave,
(d) sine wave.

(5) The main component in the smoothing stage of a power supply would be a:

(a) large value capacitor,
(b) low value capacitor,
(c) zener diode,
(d) variable resistor.

(6) The timebase circuit of an oscilloscope produces a waveform which takes the form of a:

(a) triangular wave,
(b) sawtoothed wave,
(c) square wave,
(d) sine wave.

(7) If a waveform repeats itself at regular intervals it is known as a:

(a) regular waveform,
(b) annular waveform,
(c) periodic waveform,
(d) repetitive waveform.

(8) A sine wave can be produced by a:

(a) relaxation type oscillator,
(b) half-wave rectifier,
(c) full-wave rectifier,
(d) harmonic type oscillator.

(9) A most useful instrument used to study waveforms produced by a circuit is called:

(a) an oscilloscope,
(b) a signal generator,
(c) a relaxation type oscillator,
(d) a multivibrator.

(10) Pulse waves are used in electronic circuits to:

(a) measure frequency,
(b) trigger thyristors,
(c) measure sound levels,
(d) test diodes.

Chapter 9
Electromechanical Components

9.1 Switches

Electromechanical switches do not belong exclusively to electronics; they are found in all sections of electrical engineering. In broad terms, a switch could take the form of a massive air-blast circuit-breaker capable of breaking several thousand amps, to a tiny PCB device capable of breaking only a few milliamps. Many purely electronic switches exist in the form of transistors, thyristors and triacs, and these have been discussed elsewhere in the book.

Whatever the type of switch, and whatever the sort of use we are going to subject it to, we will require certain basic information about it before we can ascertain whether it will be suitable for our use.

The contact configuration of switches

Basically this is what the switch will do when it is operated. All switches do the same job: they either make or break a connection between two or more contacts, thus permitting or preventing current flowing in the circuit. When switches are closed, allowing current to flow, they are said to be made; when open, and preventing current from flowing, they are said to be broken . Switch descriptions include a number of other essential items of information including poles and throws.

Poles This refers to the number of different sets of switch contacts that may be broken when the switch is operated. The single pole switch (SP), for example, has the ability to break one set of contacts when operated, while the double pole (DP) and triple pole (TP) switches will operate two and three sets of contacts respectively.

Throws This refers to the number of contacts that can be switched between. A single throw switch will be switched across one set of contacts as in a one-way lighting switch, while a double throw switch has a common terminal which can switch between either of two contacts, as in a two-way lighting switch. Switches, especially the rotary type, are available with many throws, perhaps a dozen or more.

Some common switch configurations are shown on Information Sheet No. 9A.

Information sheet No.9.A.
Switch configurations

Single pole-
Single throw
s.p.s.t.

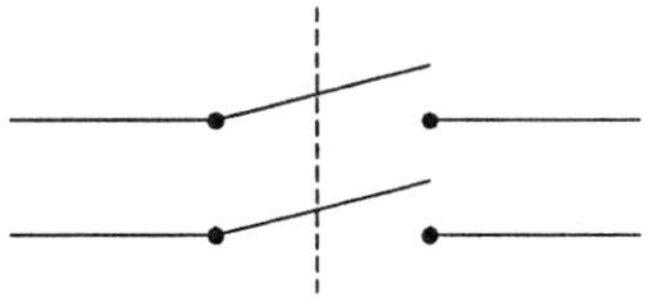

Double pole-
Single throw
d.p.s.t.

Single pole-
Double throw
s.p.d.t.

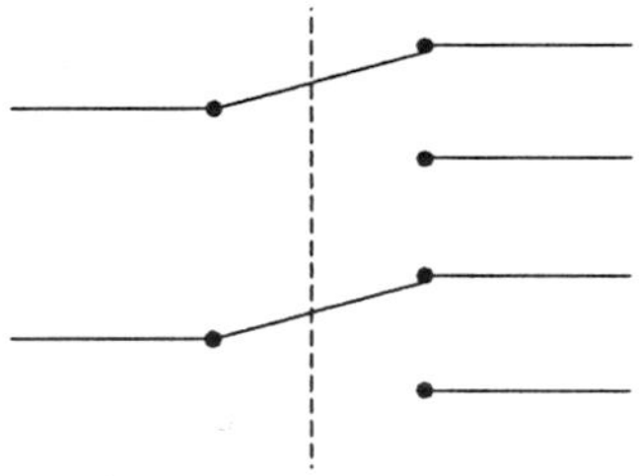

Double pole-
Double throw
d.p.d.t.

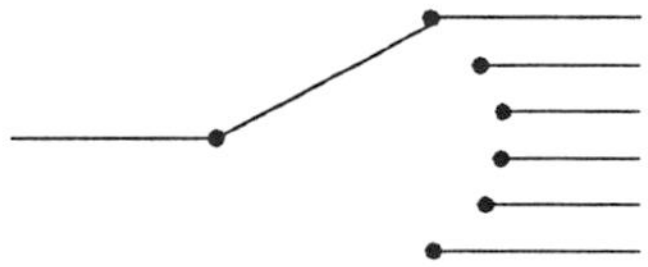

Multipole

Push to break

Push to make

The rating of switches

(1) *The current rating of switches*. This is the maximum prospective current that may be made, or the maximum steady current which may be interrupted, with a purely resistive load. It is the usual rating that is quoted for all types of switches and relays, although the power rating may in certain circumstances be given as well. Switches intended to make or break highly inductive or highly capacitive loads may have to be derated.
(2) *The current-carrying capacity*. This is different; it is the steady current that the switch will carry when its contacts are made. It is limited by the heating effect (I^2R) and is generally greater than (1) above.
(3) *The voltage rating*. This is the maximum working voltage and is based upon such considerations as contact separation, degree of separation between poles, insulation materials and safety aspects.

Switch action

This is basically how the switch achieves the contact described above.

- *Normal switch action* occurs when a switch is operated and its state changes, say from *off* to *on*. To change its state back again requires the switch to be operated once more. This action applies to the majority of manually operated switches.
- *Momentary action* occurs when a push button switch is depressed to change its state from *off* to *on* and this state only remains whilst the operating force is applied. A bell push is a good example of this type of action.
- *Latching action* is when a switch is operated and its state changes from *off* to *on* or *on* to *off* and will not return to the original position once the operating force has been removed, but remains at the lower position until a further operating force is applied. This is not the same as normal switch action, as the switch (more usually a push button) is retained in position either by mechanical means or even by a magnetic device. A good example of a latching action switch is the mushroom-headed emergency stop (turn to release) button.

9.2 Electrically operated switches

Variously named solenoids, contactors or magnetically operated switches, the relay is quite simply an electrically operated switch. Its current rating, voltage rating, current-carrying capacity and contact configuration are the same in every respect to that of the switch. It is its method of operation that distinguishes it from its more lowly counterpart.

All electrically operated switches have one thing in common: their switching operation, of whatever sort, is evoked by electromagnetic attraction, i.e. current is supplied to the coil of an electromagnet and this operates the device.

Relay

When the coil in a relay is energized, it will close or open a set (or sets) of contacts. The configurations are exactly the same as for the switch; however, the contacts are referred to as being normally open (NO), normally closed (NC), or change-over (CO). The initials are used quite a lot and should be learnt. They are often combined with other initials that you are already familiar with. For example, a SPNO would be a single pole normally open switch, while a 1A DPCO would be a relay whose maximum switched current rating was one amp and had a double pole changeover switch action.

The coils can be light duty or heavy duty and care must be taken to ensure that the control circuit current feeding the coil is well within the coil's rating.

Like any other switch, the current rating, voltage rating and current-carrying capacity of the contacts should be suitable for the purpose for which the switch is required. Another important detail, especially with relays that are to be used in conjunction with other electronic components, is the time that it takes to operate and the time that it takes to release.

Latching relays

Latching relays come in all shapes and sizes and there are some very clever methods of achieving the latching process. There is a type which uses two coils, one to set and one to reset. Another type, which is PCB-mounted needs only an impulse of a fraction of a second to set it. This pulse gives the ferrous core enough remnant magnetism to hold the contacts in the set position. A series diode is used for a.c. control, and the device can be reset using a similar pulse with reversed connections. Yet another type uses two coils as before but has a permanent magnet to assist the holding of the contacts in position.

Solenoids

In broad terms the solenoid differs from the relay in that, for the most part, the electromagnetic action operates some armature, lever, air or water valve, striking arm or plunger. These may ultimately cause switch action to take place, but not as a direct consequence of the current being applied. Examples of the use of solenoids are door bells, air and water valves, and electrically operated door latches.

Contactors

These are for the most part used for heavy-duty applications. Here the contacts will be designed to switch heavy currents such as lighting- or power-circuits, while the coil will be connected to a control circuit which will be of a much lighter current and/or voltage.

9.3 Sockets and plugs

There are far too many different types of sockets and plugs to mention individually. The following are the more popular types:

- *IEC mains plug and socket* – the familiar flat, three-pin plug (not the 13 amp one) with wide European acceptance. Rated at 250 V a.c. 6 A, it finds popular use with radios, portable televisions and the like.
- *Two-pin American* – a flat, two-pin plug designed to fit the standard American two-pin socket. It is rated at 250 V a.c. 10 A and should only be used for domestic equipment when the plug and socket meet the appropriate British Standard (see IEE Wiring Regulation 553-01).
- *Two-pin moulded plastic* – much used on tape cassette recorders, calculators, etc. This tough plastic plug is moulded onto a 1.5 m lead.
- *Jack plugs* – available in numerous guises and not suitable for mains voltage connections. They find use with microphones, headphones, radio aerials, etc. Popular diameters are 2.1 mm, 2.5 mm and 3.5 mm. Although having only one pin, they achieve multiple connection by the clever isolation of the different sections with insulation.
- *Coaxial plugs* – used almost exclusively for television aerial connections; there are various hybrids used on video recorders, etc. These plugs are designed specially to terminate the coaxial cable with the outer conductor taking the form of a woven copper braiding surrounding the insulated inner core.
- *'D' connectors and plugs* – designed for use with the flat ribbon cable or circular multicore double-screened data cable. These are in common use for such things as computer serial and parallel connections to printers, disk drives, VDUs, datacomms, gender benders, etc. Offering locking action with jack screws or spring clip retainers, these popular connectors come in both male and female socket configurations.
- *'DIN' plugs and sockets* – ranging from three-way to eight-way configuration, these versatile connectors are rated at 100 V a.c./150 V d.c. at 2 A, and are much used for mixers, amplifiers and professional standard audio equipment interconnections generally.

9.4 Edge connectors

Many electronic circuits are now designed on a modular system, that is to say each stage of the system is completed on its own PCB board. The advantages of this are fairly obvious: an engineer need not spend a lot of *down time* repairing the equipment but simply exchanges the faulty board for a new one. This type of

Information sheet No.9.B.
Various common connectors

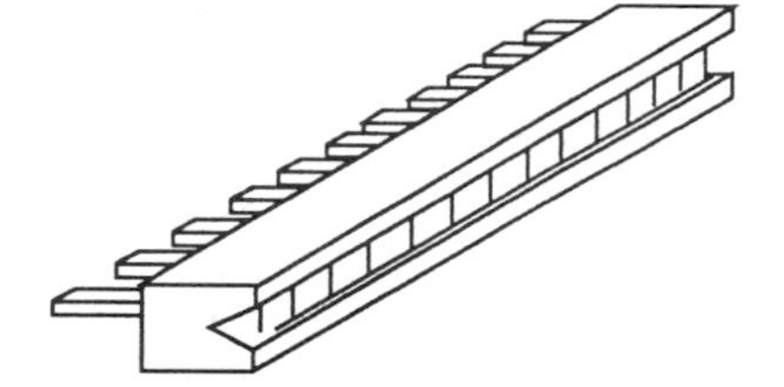

Edge connector

B.N.C. connector

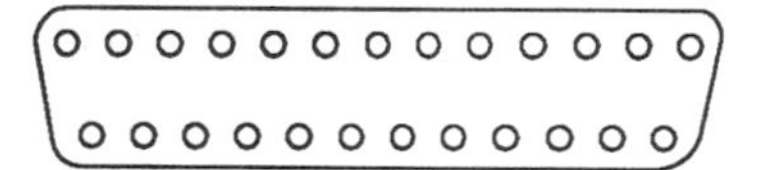

D-type connector

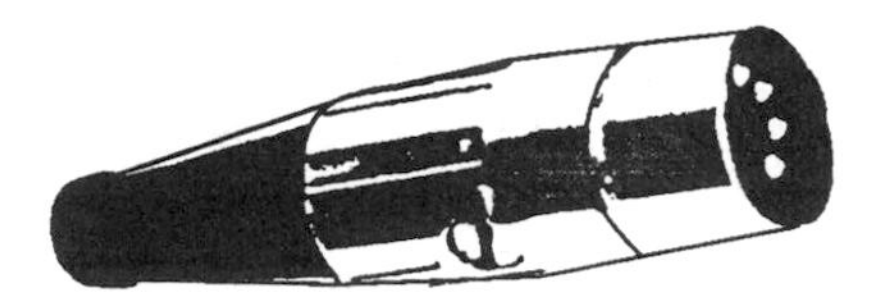

DIN connector

operation has been made even easier in recent years by the use of the edge connector.

The circuit board in question will not be connected to the other stages of the system by cabling, but will have an edge connector along one side. The circuit on the board will have all its connections taken to a number of flat, gold-plated connection pins no thicker than the actual board and spaced out from each other like the teeth of a comb. A typical standard board will be made of epoxy glass resin measuring approximately 203 × 114 × 1.6 mm, and have a 43-way gold-plated edge connector. The board will have a slot at position 37 which prevents the board being inserted the wrong way round.

The board is then plugged into an edge connector socket, which in turn is usually mounted in a special rack arrangement. There are a number of popular standards in use: IBM, Apple Mackintosh, DIN, Eurocard, etc. The idea is that the faulty board is pulled out and replaced with a new one; this is accomplished in only a few minutes and the client has the use of the equipment back in a very short space of time. The engineer can then trace the fault on the board and replace any components that are found faulty at leisure. In recent times the cost of the boards has come down so much that very often they are simply thrown away, the cost of the engineer's time far outweighing the cost of the board.

9.5 Insulation displacement connectors (IDCs)

Many of the connector sockets have plated pins to which ribbon cables can be soldered. In recent years a form of connection which dispenses with the need to carry out the soldering of these connections has evolved. These ribbon cable connectors have cable piercing tines made from nickel-plated phosphor bronze. These effect a permanent four-point contact with each individual conductor of the ribbon cable via 'V'-shaped teeth. A simple tool for facilitating the termination of IDCs can be obtained for use in an engineer's vice. With the connector and cable positioned in the tool, it is placed in the vice and the vice tightened. This forces the connector tines into the cable, thereby piercing the outer insulation and making connection to the inner conductors. More sophisticated tools are available, as are tools suitable for mass production. Information Sheet No. 9B gives examples of some types of connector. Recently, manufacturers of some types of telephone outlets have adopted the IDC method of termination and have included a disposable tool with the accessory.

SUMMARY

This chapter has described electromechanical components associated with electronic circuitry. The following important points should have been learned:

- That electricians must be able to recognize and identify the more popular types of switch;
- The meaning of the terms *pole*, *throw* and *contact configuration*, and how they apply to switches used in electronic equipment;
- The nature of switch action;
- What is meant by the current rating, current-carrying capacity and voltage rating of switches and the importance of this knowledge when ordering switches from a supplier;
- The operation of relays, latching relays, solenoids and contactors;
- That electricians must be able to recognize the more popular types of plugs, sockets and connectors;
- The purpose of edge connectors;
- The meaning of IDC (insulation displacement connector), and how this type of connection is achieved.

Test 9

Choose which of the four answers is the correct one.

(1) The steady current carried by a closed switch is its:

(a) current rating,
(b) current-carrying capacity,
(c) power rating,
(d) voltage rating.

(2) When a switch is closed the contacts are said to be:

(a) selected,
(b) shut,
(c) made,
(d) broken.

(3) When a switch is given the initials SPNO it is said to be:

(a) switch position not operating,
(b) single pole not operating,
(c) switch position normally open,
(d) single pole normally open.

(4) To speed up the replacement of circuit boards they often have:

(a) quick release catches,
(b) edge connectors,
(c) self tapping screws,
(d) parts numbers.

(5) The separate sets of contacts operated by a switch are called:

(a) the number of poles,
(b) the number of ways,
(c) the number of gangs,
(d) the contact number.

(6) Ribbon cables are terminated with a connector called:

(a) an insulation displacement connector,
(b) a flat based connector,
(c) an edge connector,
(d) a block connector.

(7) The principle of operation of a relay is by:

(a) high frequency induction,
(b) capacitive reactance,
(c) electromagnetic attraction,
(d) inductive reactance.

(8) The voltage rating of a switch does not depend upon:

(a) contact separation,
(b) separation between poles,
(c) insulating material,
(d) number of poles.

(9) A contactor is different from a relay in that it usually:

(a) has a d.c. coil,
(b) has an a.c. coil,
(c) makes or breaks heavy currents,
(d) makes or breaks light currents.

(10) A solenoid differs from a relay in that it does not have:

(a) an operating coil,
(b) direct switch action,
(c) heavy insulation,
(d) light insulation.

Chapter 10
Safety and Testing

10.1 The Electricity at Work Regulations 1989

The Electricity At Work Act 1989 stresses that it is always preferable, from the point of view of safety, that work on or near electrical equipment should be carried out when the equipment is dead. Regulation 14 does, however, recognize that there are circumstances where it is unreasonable, having regard to all the relevant factors, for equipment to be dead while work is being carried out. The carrying out of tests on live electronic equipment could be considered more reasonable than live working though the basic safety rules still apply:

- Only uncover as much work as is really necessary;
- Use the correct approved testing equipment and tools;
- Keep any sections exposed under supervision.

In most circumstances it would be difficult to test *dead* electronic equipment. However, where regular live testing is carried out – in TV repair shops, appliance testing sections or even units produced in a manufacturing process – serious consideration should now be given to the establishment of a designated test area in which special precautions have been taken. For example, in TV repair workshops isolating transformers can be used to supply the bench sockets. These will help to create an earth-free situation – very important when you consider that the TV chasses are often above earth potential. If the circuit is backed up by an RCD so much the better. There are many other areas where these precautions should be taken such as laboratories, schools, etc., and details of publications on electrical safety for some of these areas are given in Appendix 1 of the Electricity At Work Regulations Memorandum.

10.2 Testing resistors

Resistors are tested by use of an ohmmeter, set on a scale suitable for the value of the resistor under test. It is worth mentioning at this point that when using a multimeter, the red terminal identifies the positive connection and the black terminal the negative connection for measuring voltages and currents only; the connections **must be reversed** when the instrument is used on the ohms scale. This is because of the way the meter movement is connected to the internal battery.

It is very unlikely that a fixed resistor would *go* short circuit. It would be more likely to go high in value or completely open circuit. Film resistors likely to go open circuit because as we know they are essentially a thin resistive spiral which may break. Faults on fixed resistors are most often caused by operating the resistors outside their parameters or by rough handling.

A variable resistor has moving parts and is therefore potentially less reliable than a fixed resistor. It depends for its operation on a good electrical contact between wiper and track. A partial failure can occur when dust or dirt finds its way under the wiper contact; this results in an increase in noise. Cleaning fluids and sprays can be very successful in combating this. Complete failure is usually due to the resistor going open circuit for all the same reasons as the fixed resistor.

10.3 Testing capacitors

Capacitors can go open circuit or short circuit, although electrolytic capacitors are likely to do the latter. Aluminium electrolytic capacitors tend to go *leaky*, allowing the charge to leak away quicker than other types; for this reason they are not recommended for timing circuits.

Sometimes it is necessary to test a capacitor to verify that it is in serviceable condition. The best method is to use a purpose-made capacitor tester, which will not only indicate whether the component is sound, but will indicate its actual capacitance.

Another method is to use a standard multimeter set on its highest resistance range; remember to take the precaution of reversing the test leads mentioned earlier when the ohms scale is selected. This is particularly important with electrolytic-type capacitors as they themselves are polarized. The meter is connected to the two leads of the capacitor, and at the same time the scale is observed. With a sound capacitor the needle of the meter will flick across the scale and then settle back into its original position. The time taken for the pointer to return to its original position will depend upon the value of the capacitor, as this time is related to the capacitor charging.

If, when testing the capacitor, the pointer was to move across the scale and stay in that position, this would indicate a short-circuit fault. If on the other hand it failed to move at all, this would indicate that the capacitor was open circuit.

10.4 Testing diodes

Diode faults are fairly straightforward in that they either go short circuit, and so behave as if they were a piece of wire, or they go open circuit and pass no current at all. The most common fault with diodes is the open circuit fault.

We learned in our earlier work that the function of the diode is to pass current in one direction and block it in the other. Therefore, when we carry out a test on a diode with an ohmmeter, it should indicate a high resistance one way and a low resistance the other.

In practice it is not necessary to know the actual diode resistance in the forward or reverse direction. Therefore, we can perform a simple go/no go test on the component. To do this, set the ohmmeter to read with a full scale deflection of say 10 kΩ. When the device is tested one way the needle should not move, indicating a high resistance. Reversing the connections should result in a full-scale deflection indicating a low resistance. If the needle did not move when connected in either direction, then this would indicate an open circuit diode. With the needle giving full scale deflection when tested in both directions this would indicate a short circuit condition. Despite the fact that the electricians' insulation tester (megger) has an ohms scale, it is not recommended that this be used to test electronic components.

10.5 Testing transistors

It is perhaps unfortunate that, along with other semiconductor devices, as the temperature of the transistor increases its resistance falls. When the resistance falls more current flows and the hotter the device becomes. This cycle continues until the high temperature causes the device to break down. This cycle is known as thermal runaway and precautions such as providing heat shunts and ensuring that there will be no overloading of the device, have to be taken when the circuit is being designed, to prevent or limit this.

Testing a transistor is fairly straightforward and can be carried out with an ohmmeter, or if a number of transistors are to be tested on a regular basis, a purpose made test circuit, as described on Information Sheet No. 10A could be constructed.

With the ohms scale selected on the multimeter and the leads reversed as mentioned above, *a sound npn transistor* will give the following readings:

Positive lead to base and negative to emitter = low reading;
Positive lead to base and negative to collector = low reading;
Negative lead to base and positive to emitter = high reading;
Negative lead to base and positive to collector = high reading.

The reading will be high between collector and emitter whichever polarity is used. Any readings that differ from the above will usually mean that the transistor is defective.

With the ohms scale selected on the multimeter and the leads reversed as mentioned above, *a sound pnp transistor* will give the following readings:

Positive lead to base and negative to emitter = high reading;
Positive lead to base and negative to collector = high reading;

Information sheet No. 10.A.

Transistor tester circuit diagram

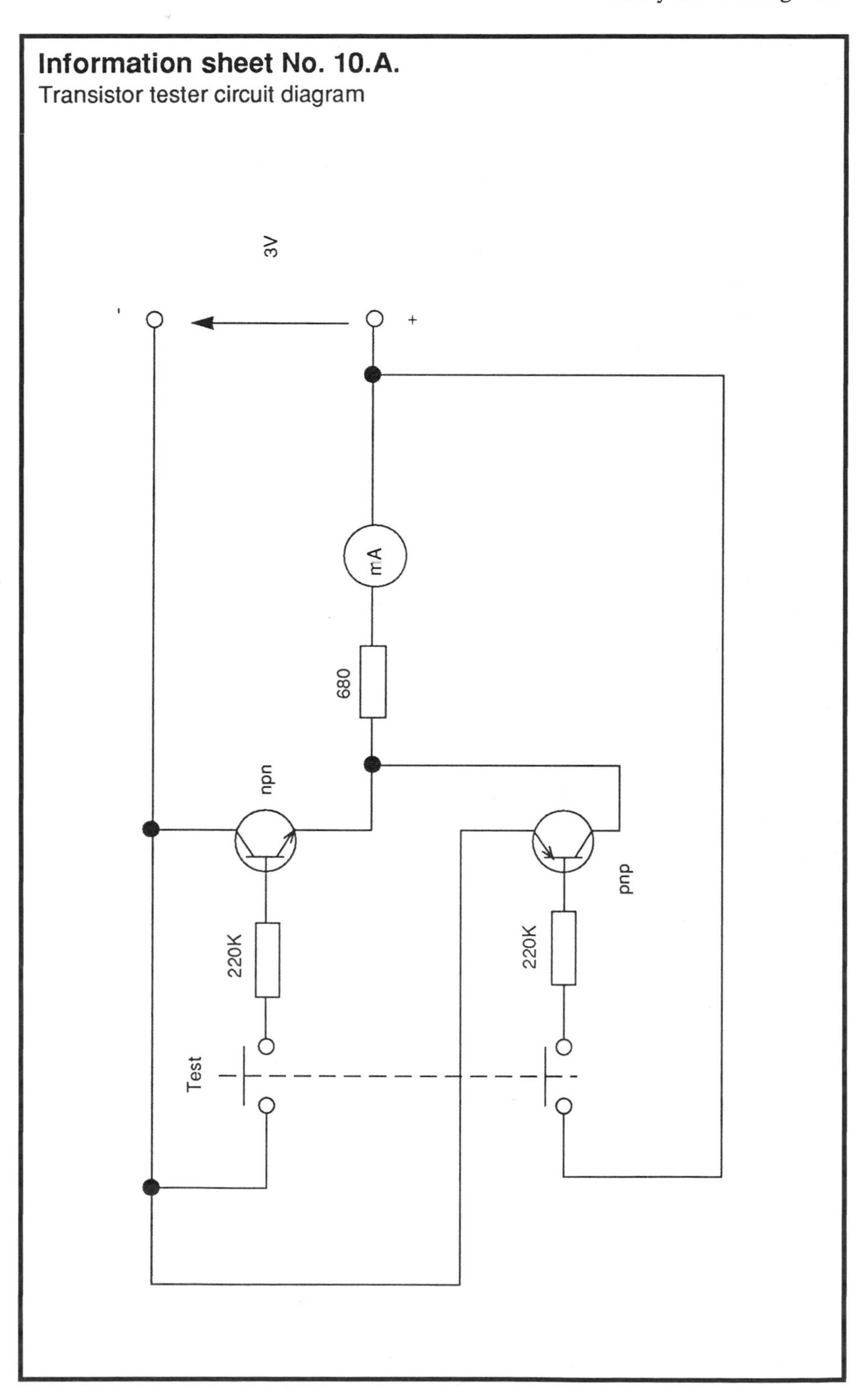

Negative lead to base and positive to emitter = low reading;
Negative lead to base and positive to collector = low reading.

The reading will be high between collector and emitter whichever polarity is used. Any readings that differ from the above will usually mean that the transistor is defective.

10.6 Testing thyristors and triacs

Thyristors and triacs are semi-conductor devices; they fail usually because the temperature inside the component has been excessive. This could be due to either overloading of the device, or to an incorrectly designed heat sink arrangement.

It is possible to carry out certain checks with the component in circuit. By injecting a trigger pulse into the gate, we can determine whether it is the trigger circuit or the thyristor that is at fault. With the current isolated, a test can easily be made to determine whether there is a short circuit between cathode and anode. In addition, because the gate to cathode effectively forms a diode, a test to verify this would give a low reading in one direction and a high reading in the other (see Section 10.4 Testing diodes).

10.7 Test and measuring instruments

Some of the equipment used by electricians such as voltmeters, ohmmeters and ammeters are used by the electronics engineer, too, and these are described in detail in Book Three of *Electrical Installation Practice*. (Under no circumstances should an insulation tester be used on electronic circuitry as serious damage will be caused.)

There are, however, a number of other instruments that are required for electronic work which may not be so familiar, and these include:

(1) Oscilloscopes,
(2) Signal generators,
(3) Logic pulsers and probes,
(4) Frequency meters,
(5) Transistor testers.

10.8 The cathode ray oscilloscope

The cathode ray oscilloscope (CRO) is a measuring instrument that finds an application in every branch of electronics. It is a most useful instrument allowing the user to view waveforms on its screen. Indeed CROs that will display the waveform of any voltage from d.c. to 30 MHz are now common, and special-

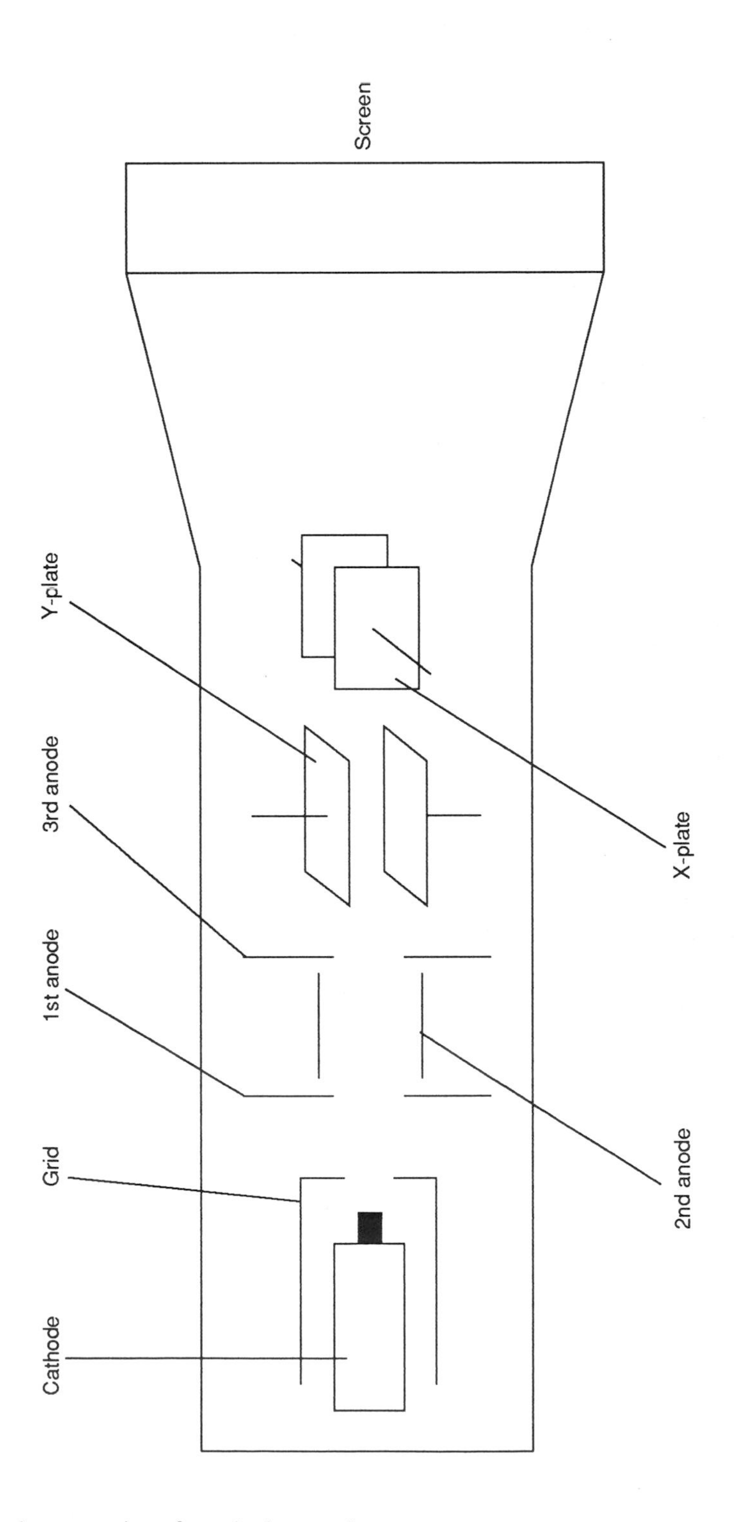

Fig. 10.1 Internal construction of a cathode ray tube.

purpose ones will respond to frequencies as high as 1000 MHz. With this facility, fault-finding in electronic circuits is substantially simplified. In fact, certain faults detected by the CRO cannot be detected by other traditional test methods. Another advantage of the CRO is that the input impedance to the instrument is extremely high (1 MΩ). This means that it takes very little current from the circuit under test, something that tends to occur with traditional moving coil instruments.

The internal construction of the oscilloscope

At the heart of the oscilloscope is the cathode ray tube (CRT), the construction of which is in four sections:

- The device which produces electrons (the electron gun);
- The focusing system;
- The deflecting system;
- The screen itself.

The *electron gun* produces a fine beam of electrons which are given off from a heated cathode by thermionic emission. These electrons move through the length of the tube at a high velocity towards the glass screen. The inside of the screen is coated with a fluorescent material that will give off light when the electrons strike it. Inside the tube is a vacuum to enable the electrons to pass down the tube without interference. A simplified diagram of a CRT is shown in Fig. 10.1.

The electrons are attracted to Anode 1 by the very high voltage on it (because unlike charges attract). On their way to Anode 1 electrons pass through a grid and by turning a knob marked *intensity* on the front of the oscilloscope this grid is made either more or less negative with respect to the cathode. This controls the number of electrons and therefore the intensity of the spot formed by the beam on the

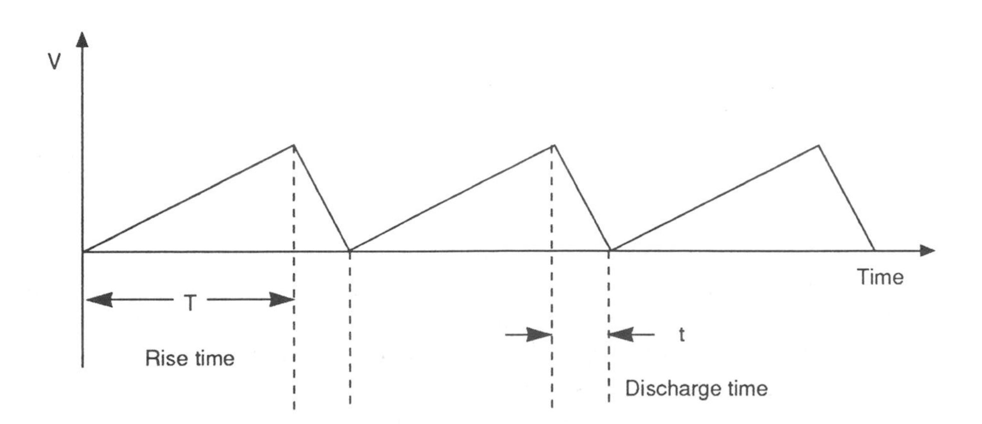

Fig. 10.2 Sawtooth waveform.

screen. Some of the electrons will be collected by Anode 1 when they reach it but others will pass through a hole in its centre.

The *focusing system* consists of Anode 2 and by operating the focus control on the front of the CRO the voltage on this anode is varied and the beam is focused to give a sharp image on the screen. Anode 3 further accelerates the beam of electrons towards the screen.

The *deflecting system* moves the beam on the screen and comprises the two X-plates, X1 and X2, and two Y-plates, Y1 and Y2. The X-plates are connected to the timebase; this moves the beam horizontally across the screen in the way described in Chapter 8.

The waveform of the timebase should be very similar to that shown in Fig. 10.2. From the diagram it can be seen that the distance travelled horizontally by the spot must be directly proportional to time so in effect the horizontal or X axis is the time axis. The Y-plates move the beam vertically across the screen so if at the same time the signal voltage is applied to the Y-plates the combined result is that the waveform for the signal concerned is traced on the screen.

The timebase frequency must be the same as that of the signal (or a submultiple of it) for correct representation of the waveform. This is adjusted by two controls on the front of the CRO: one, called the time control, selects the approximate frequency, while the other, known as the fine frequency control, adjusts the oscillator frequency to the desired value. The electrons, having carried out their task, return to the supply via a terminal on the tube.

The *screen* is covered internally with a fluorescent coating which emits light when hit by the electron beam. The screen colour depends on the type of coating and can be green, orange or blue, with green being most popular.

Although the waveform on the screen is made up of a rapidly moving trace it looks continuous to the human eye by what is called persistence of vision. The eye cannot detect the rapid changes that are involved in the display.

Using the oscilloscope

The screen is a cathode ray tube much like a miniature television, which has a grid superimposed upon it. This consists of horizontal and vertical lines 1 cm apart and is referred to as the graticule. The horizontal lines form the time axis (sometimes referred to as the X axis) and the vertical lines form the voltage axis (sometimes called the Y axis). Information Sheet No. 10B shows a typical layout of the front of an oscilloscope

The timescale of the horizontal axis is selected by a rotary switch; this selection determines the time value of each horizontal centimetre. For example, if 2 ms has been selected then each horizontal square will represent 2 ms. If a dual beam oscilloscope is being used the time selected applies to both of the beams.

The voltage scale of the vertical axis is again selected by a rotary switch; the voltage selected determines the value of each vertical square. For example, if 5 V has been selected then each vertical square will represent 5 V. If a dual beam

Information sheet No.10.B.

Typical oscilloscope layout

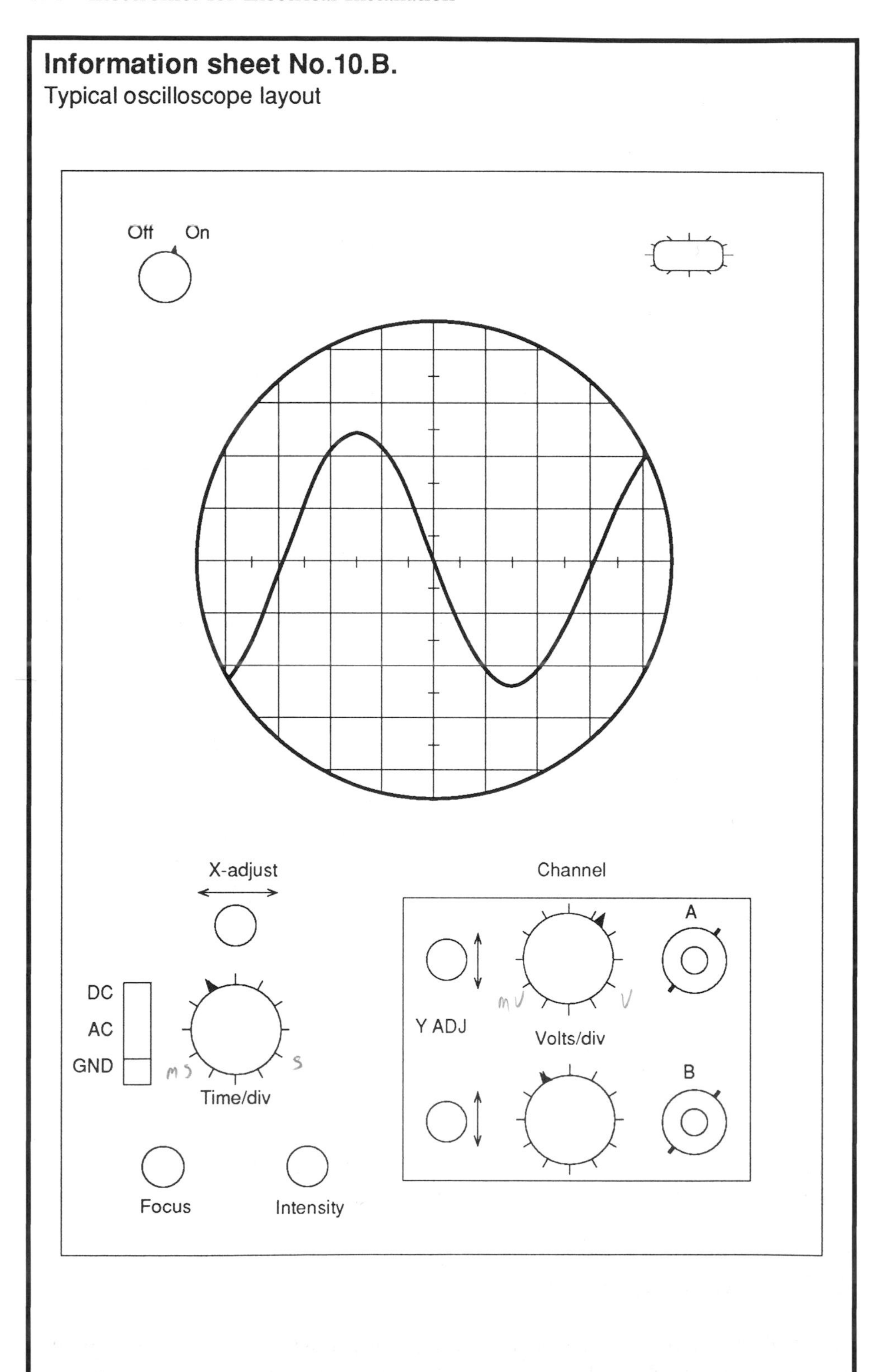

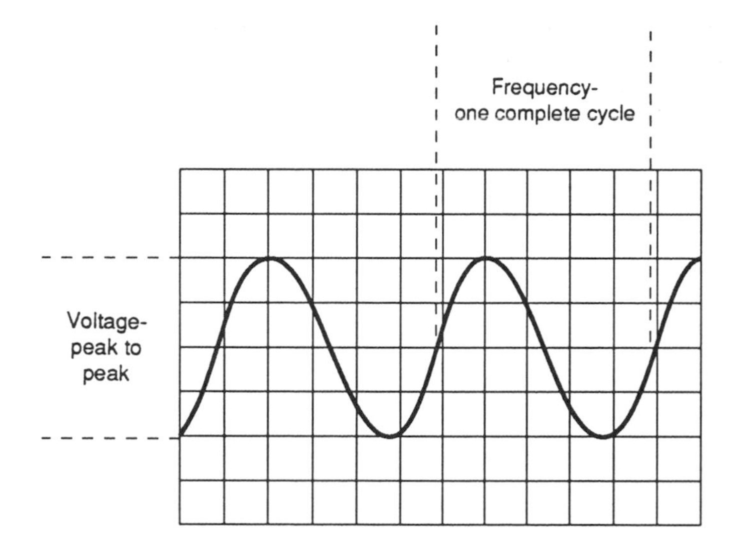

Fig. 10.3 CRO trace.

oscilloscope is being used then each channel has its own independent voltage selection switch. By correctly selecting the time and voltage scales the optimum waveform can be displayed.

With the signal under test connected to the Y-input and the controls adjusted as described above we can now measure the voltage and the frequency. Figure 10.3 shows a typical trace and if we use the figures selected above we can work out the values represented by the trace.

By counting the number of centimetre squares from peak to peak and multiplying this by the 5 V/cm we can establish the peak-to-peak voltage. In our example this would be $4 \times 5 = 20$ V. This will be divided by 2 to obtain the peak voltage of 10 V. By further multiplying this by a factor of 0.7071 we can arrive at a rms voltage of 7.071 V.

By counting the number of centimetre squares horizontally for one complete cycle and multiplying this by the 2 ms/cm we can establish the time. In our example this would be $5 \times 2 = 10$ ms. From this we can use the following formula to work out the frequency:

$$f = \frac{1}{T} = \frac{1}{10 \times 0.001} = 100 \text{ Hz}$$

From the above we can say that the waveform represented in Fig. 10.3 has a voltage of 7.071 V at a frequency of 100 Hz.

10.9 The signal generator

A signal generator (also called a function generator) is a very useful piece of equipment that finds applications in fault-finding, circuit-testing and measuring.

Information sheet No.10.C.

Typical signal generator

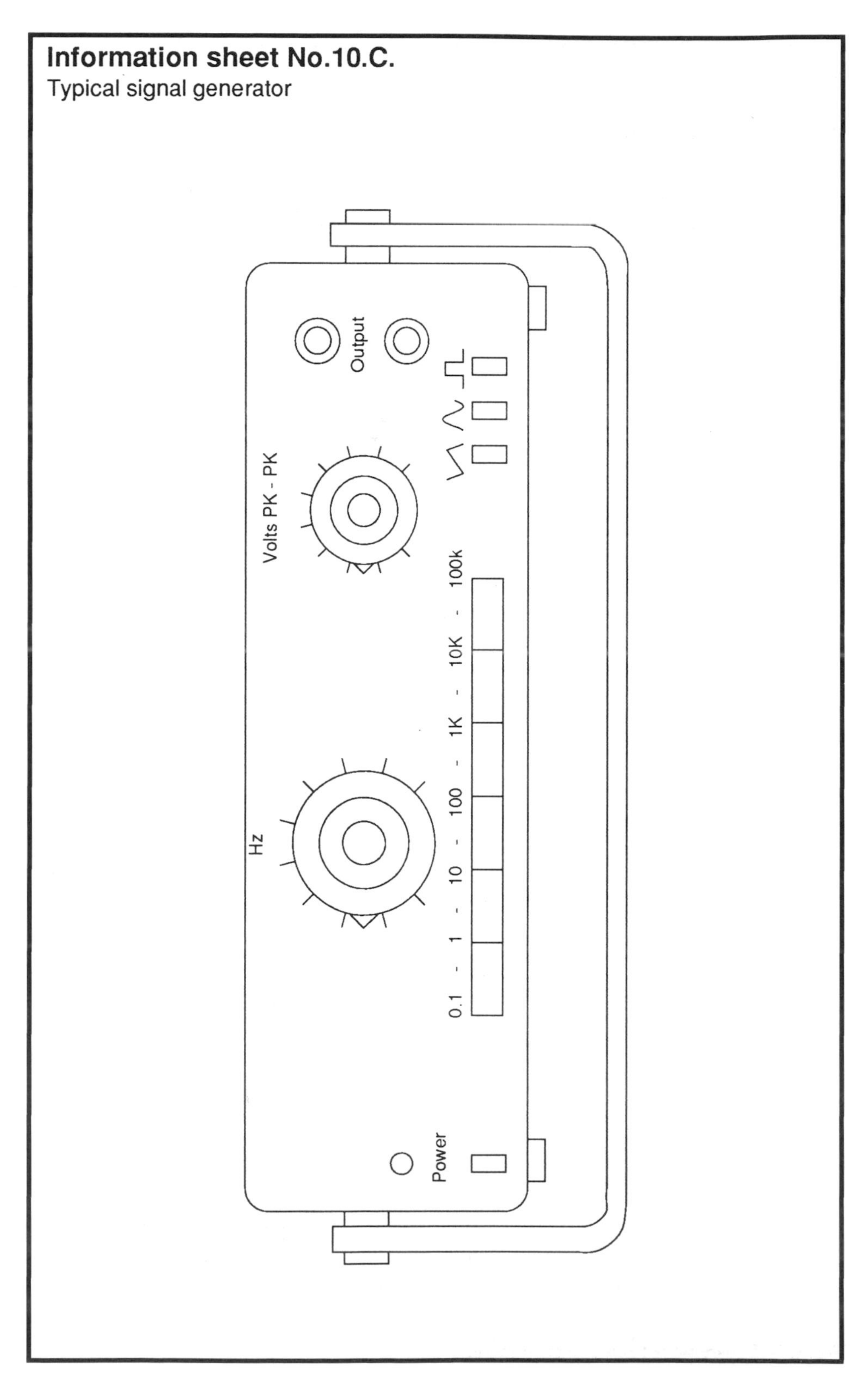

A signal generator is superior to a simple oscillator circuit because of the following features:

- The output frequency is adjustable, from very low frequencies, say 0.5 Hz (1 cycle taking two seconds) to high frequencies, well over 1 MHz;
- The peak-to-peak voltage is fully adjustable, from a few millivolts to over 10 V;
- Several output waveforms may be generated within the unit including sine waves, square waves, and triangular waves.

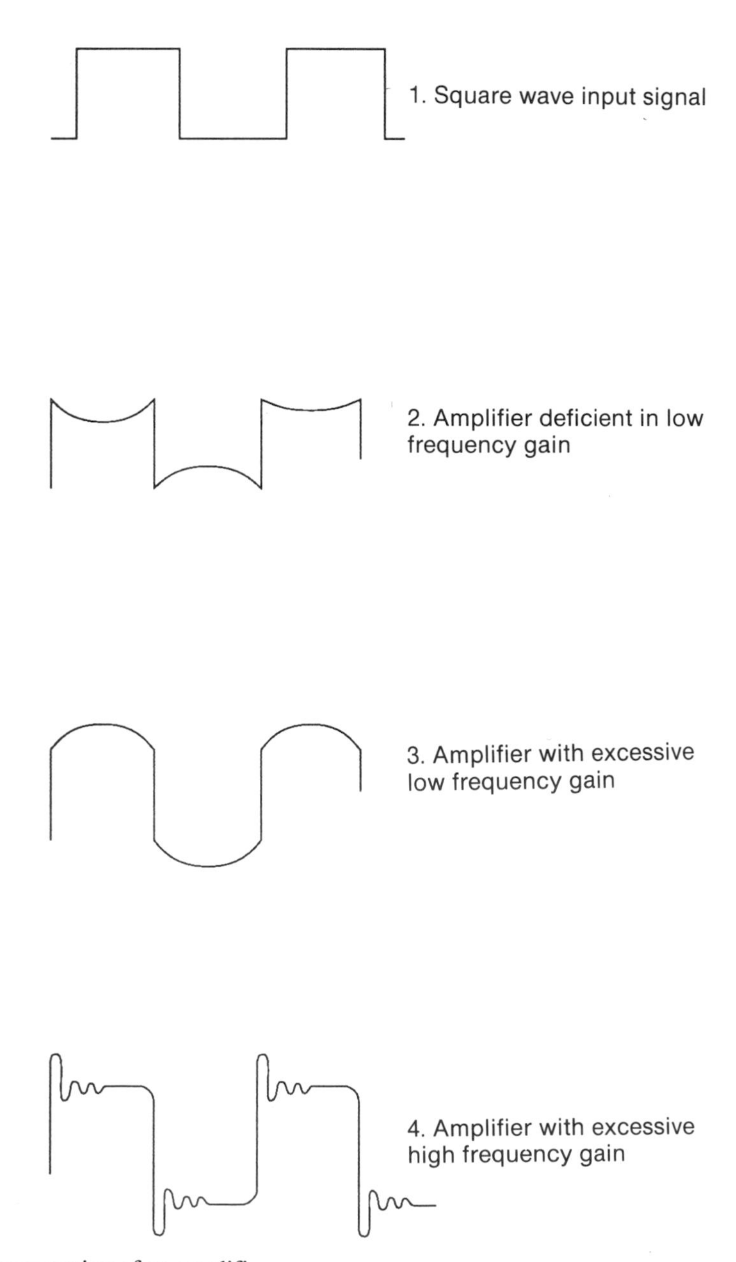

Fig. 10.4 Square wave testing of an amplifier.

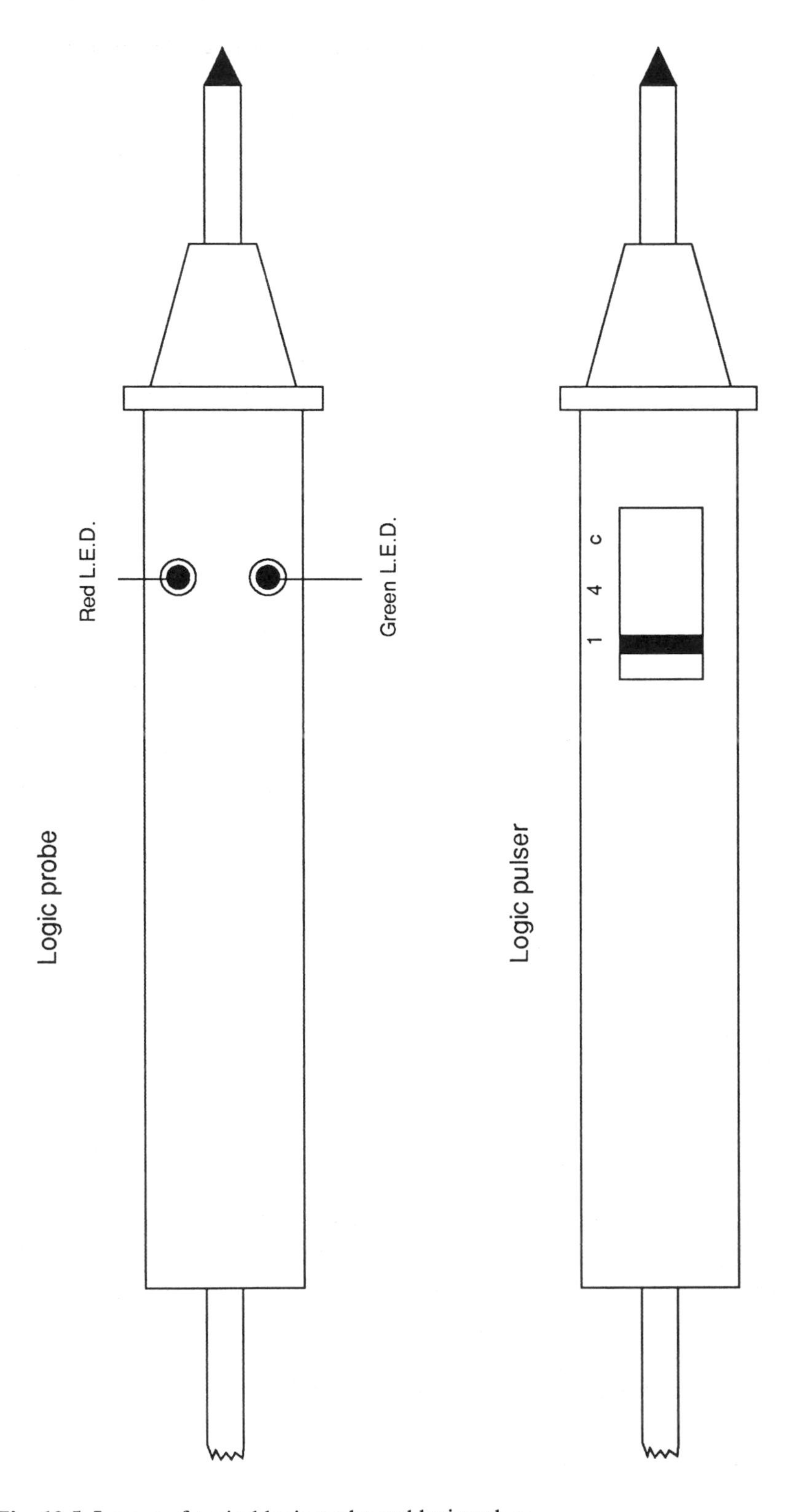

Fig. 10.5 Layout of typical logic probe and logic pulser.

A typical signal generator is shown on Information Sheet No. 10C.

A good example of its use is in the testing of an amplifier circuit. The response of an amplifier with regard to frequency and distortion may be assessed by feeding a square wave in at the input and displaying the output waveforms on an oscilloscope. By varying the frequency, the frequency response of the amplifier may be assessed. Further, the amount of distortion the amplifier introduces can be seen.

A square wave of 1 V is fed in; naturally, because the circuit is an amplifier, the output will be larger, but ideally it should still be an accurate square wave if the amplifier performs well. If it is not then the circuit is introducing distortion. Good and poor outputs from an amplifier are shown for a square wave input in Fig. 10.4.

10.10 Logic pulsers and probes

The logic pulser is another piece of equipment found only in electronics; it considerably simplifies fault finding in digital circuits and systems, enabling logic gates, flip flops and counting circuits to be tested more easily.

A good logic pulser will have a tristate output; this means that it sends out pulses which are high or low, but when it is not operating it offers a high impedance to the circuit to which it is connected. This means that the logic pulser can be left connected without interfering with the logic circuit under test. A switch selects the number of output pulses from the tester, which is usually a square wave with a frequency of 1 kHz. A typical logic pulser is shown in Fig. 10.5.

Logic pulsers work from a range of voltages, typically 8–18 V; this makes them ideal for both TTL and CMOS circuits. The supply leads are reverse polarity protected and can have a variety of different hooks, insulated tips and IC test adapters fitted.

To use the logic pulser the tip is placed upon the gate or circuit input, the type of output from the pulser is set and the *on* button pressed. The pulser automatically sends out the amount of pulses set by the switch and drives the circuit, thus enabling suspect gates to be checked. The results of the test can be detected using a logic probe, or displayed on an oscilloscope.

The added beauty of this method of testing is that no desoldering of components takes place.

Logic probes are high quality diagnostic tools allowing logic states in TTL and DTL circuits to be displayed on a pin-by-pin basis. They are designed to display the output from a particular part of the circuit or system, whether permanently high or low, or if it consists of 0 and 1 pulses.

The display may indicate *H* for high, *L* for low or *P* for pulsing. Commercial designs use a coded LED output, as shown in Table 10.1.

Table 10.1 Display output of a logic probe.

Probe Display	*Logic State at Probe Tip*
Both the green and red off	Open circuit
Green on constantly	Logic 0
Red on constantly	Logic 1
Green and red flashing	0 and 1 pulses less than 10 Hz
Green and red on constantly	0 and 1 pulses greater than 10 Hz

Note: for those with short memories TTL stands for transistor-transistor logic and DTL for diode-transistor logic.

10.11 Frequency meters

It is sometimes necessary, when dealing with electronic circuits, to determine the frequency of the signal we are dealing with. This frequency (which is measured in hertz) can be defined as how many cycles of the waveform there are in one second. Two main methods of determining the frequency are available to us: one, the oscilloscope, we have discussed already; and the other, a more accurate method, is to use a purpose-made frequency meter.

Hand-held battery-powered frequency meters, quite adequate for general purposes, will measure frequencies in the range: 20–200 MHz. Typical instruments have two scales:

- Scale A will measure from 20 Hz to 10 MHz in four ranges selected by a switch;
- Scale B will measure from 5 MHz to 200 MHz, again in four ranges.

A typical hand-held meter is shown on Information Sheet No. 10D.

A considerable number of other models are available; these range from transportable designs to bench mounting types and will cover almost any range of frequencies from 0.02 Hz to 1 GHz (i.e. 1000 MHz).

10.12 Transistor testers

How to test transistors using an ohmmeter was discussed in Section 10.5, and the circuit for building your own transistor tester is shown on Information Sheet No. 10A. There are, however, as you would expect, a number of purpose-built ones on the market. They have the advantage of being able to be used in circuit and so avoid the need for any desoldering activities. Two test probes are connected across the junction to be checked and instantly two LEDs in the probes show its state.

Information sheet No.10.D.
Hand-held frequency meter

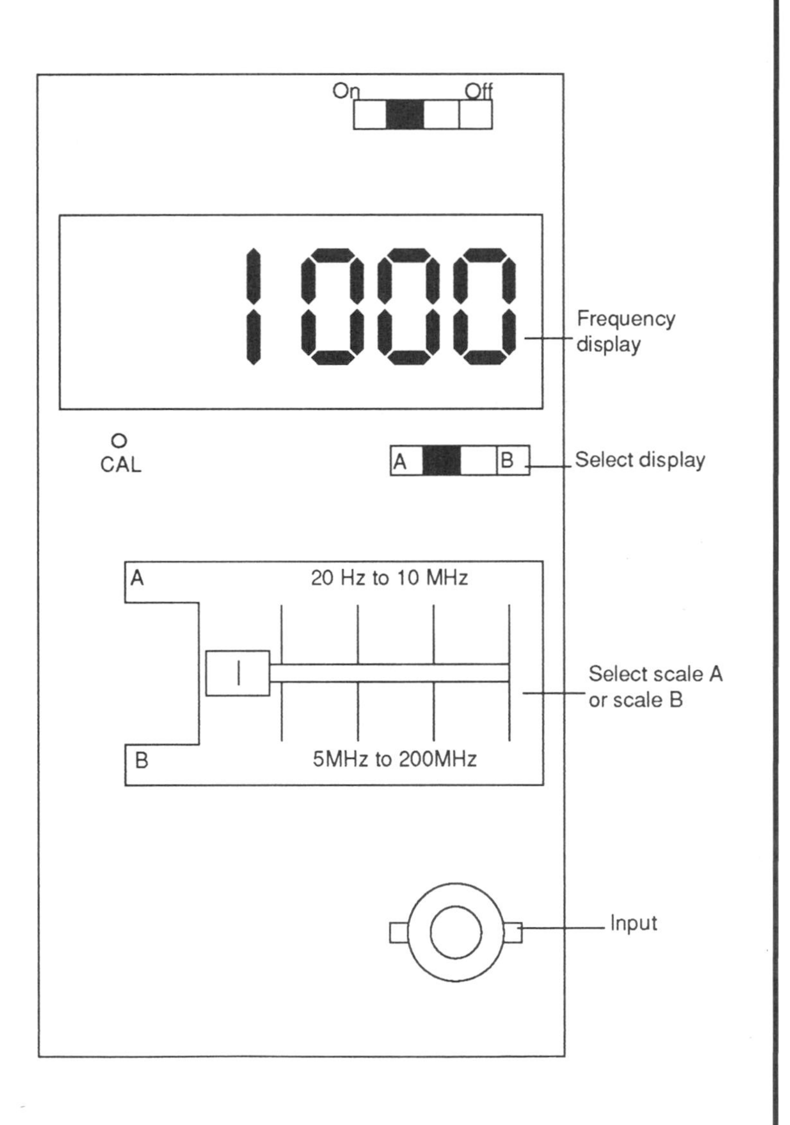

They will indicate whether the junction is open circuit or short circuit for faulty junctions and the polarity of a good junction. They are suitable for checking both pnp and npn transistors, as well as diodes and zener diodes.

SUMMARY

This chapter has described the testing of components associated with electronic circuitry. The following important points should have been learned:

- That electricians must be aware of the requirements of the Electricity At Work Regulations 1989 as they apply to the testing of equipment;
- How to carry out tests on resistors, capacitors, diodes, transistors, thyristors and triacs;
- How to use oscilloscopes, signal generators, logic pulsers, logic probes, frequency meters and transistor testers;
- The dangers associated with using an insulation tester on circuits containing electronic devices.

Test 10

Choose which of the four answers is the correct one.

(1) The values of resistors can be found by use of:

(a) an ammeter,
(b) a voltmeter,
(c) an ohmmeter,
(d) a wattmeter.

(2) The leads of a multimeter must be reversed when using it as:

(a) an ammeter,
(b) a voltmeter,
(c) an ohmmeter,
(d) a wattmeter.

(3) The safe use of electricity at work is covered by the:

(a) IEE Wiring Regulations,
(b) Electricity at Work Regulations 1989,
(c) Supply Regulations 1988,
(d) The Factories Act.

(4) A variable resistor is noisy when turned; this may be due to:

(a) the resistance value being too low,
(b) the resistance value being too high,
(c) it being a wire wound resistor,
(d) dirt under the wiper contact.

(5) The correct test results when testing a diode are:

(a) high resistance one way, low resistance the other,
(b) high resistance in both directions,
(c) low resistance in both directions,
(d) no resistance in either direction.

(6) Thyristors and diacs fail largely because:

(a) the internal temperature becomes excessive,
(b) too much current has been applied to the gate,
(c) they have not been turned on long enough,
(d) of their poor switching speeds.

(7) Tests on npn transistors show low ohmic readings when the:

(a) – lead is on the base and the + lead on the emitter,
(b) + lead is on the base and the – lead on the emitter,
(c) – lead is on the base and the + lead on the collector,
(d) + lead is on the emitter and the – lead on the collector.

(8) Tests on pnp transistors show low ohmic readings when the:

(a) – lead is on the base and the + lead on the emitter,
(b) + lead is on the base and the – lead on the emitter,
(c) + lead is on the base and the – lead on the collector,
(d) – lead is on the emitter and the + lead on the collector.

(9) The most useful instruments for testing logic gates are:

(a) voltmeters and ammeters,
(b) ohmmeters and wattmeters,
(c) frequency meters and signal generators,
(d) logic pulsers and probes.

(10) A piece of equipment used to produce waveforms for testing is a:

(a) multimeter,
(b) frequency meter,
(c) signal generator,
(d) potentiometer.

Index